Climate Change and Economics

S. Niggol Seo

Climate Change and Economics

Engaging with Future Generations
with Action Plans

S. Niggol Seo
Muaebak Institute of Global Warming Studies
Gwanak-gu, Seoul, Korea (Republic of)

ISBN 978-3-030-66679-8 ISBN 978-3-030-66680-4 (eBook)
https://doi.org/10.1007/978-3-030-66680-4

Cover design by eStudio Calamar

This Palgrave Macmillan imprint is published by the registered company Springer Nature Switzerland AG
The registered company address is: Gewerbestrasse 11, 6330 Cham, Switzerland

This book is dedicated to Monsoo.

PREFACE

The year 2019 in the timeline of global climate challenges and anthropogenic responses was saliently marked by the emergence of youth climate protests. The School Strike for Climate organized a world-wide protest of school children in March, May, and September, respectively. The September strike occurred concurrently in "150 countries" and drew millions of protesters. Across the Atlantic, the Green New Deal was proposed as a House Resolution by the youngest woman ever elected to the US Congress. On the opposite side of the Planet, as the year 2020 dawned, there was already a whisper of a major deadly virus in Chinese cities, which would turn out over the course of the year to be a once-in-a-century pandemic, the novel coronavirus.

The rising tide of youth participation in climate change debates is at the background of this book entitled "Climate Change and Economics: Engaging with Future Generations with Action Plans." The motivation of this book is to appreciate the youth movement on the Planet's climate problems but also to present to the young generations a big picture view of the climate dialogues which is encompassing of sciences, economics, and policy decisions. The present author hopes to engage young scholars and activists with a big mosaic of action plans and strategies.

Witnessing the unfolding of the year 2019, it dawned on the present author that there will be no better time to write a book-length exposition on the economic aspects of climate change and future generations. The topic is very often at the heart of the perception of the general public on climate change challenges. Notwithstanding, although the topic has long been recognized as a key economics issue to be dealt with, the research on this topic remains rudimentary and in an abstract concept, for example, through a discounting rate on future consumption. There are many dimensions on this question as well as multifarious illuminating stories to tell which researchers have neglected to explain, which I hope to get across to the readers of this book.

Departing from the currently available and well-established books on climate change and climate economics, this book is structured with a series of points of contention in the literature. Each chapter tackles one major point of contention while the entire book is structured to cover all major contentions on climate change comprehensively. The following chapters are in the book: An introduction to young climate activism (Chapter 1); farm animals in Sub-Saharan Africa (Chapter 2); Latin American rainforests (Chapter 3); Indian monsoon and agriculture (Chapter 4); cyclones, hurricanes, and typhoons (Chapter 5); the Planet's sublime grasslands (Chapter 6); energy revolution and hydroelectric dams (Chapter 7); backstop technologies (Chapter 8); polar bears and biodiversity loss (Chapter 9); coral reefs, fisheries, and mass extinction of species (Chapter 10); infectious diseases and pandemics (Chapter 11); climate negotiations for a global big deal (Chapter 12); inter-generational gaps and burden sharing (Chapter 13).

Taking note of the young readers and activists, this book is written, in a practical sense, as a Planetwide mosaic of action plans needed for the Planet's climate system and for addressing the challenges arising from a future climatic shift. In each chapter, readers will be able to draw out a list of actions and measures that are needed or should be evaluated in dealing with the problem that the chapter is focused on tackling. This book has a hands-on approach, similar to the spirit of climate protests, instead of an intricate theoretical assessment.

Deviating from the other notable books in the literature, I take a personal storytelling approach in this book. This means that many of the stories in the book are from the personal experiences of the present author. I must tell that the present author had the good luck to work in all six continents of the Planet working on one climate problem to another for the past 20 years: North America, South America, Africa, Europe, Oceania, South Asia, and East Asia. During the two decades, I have endeavored to experience many famed climate events and happenings as closely and personally as I can.

But, more critically, I mean by the storytelling approach that this book is written from the perspectives and experiences of individual citizens across the Planet with regard to manifold unique climatic events they experience. In part, this approach was motivated by the present author's desire to write the book in a manner that appeals most to the young readers.

This book is written for the young readers as primary audience, specifically, the gen-Zers. They may be a college student, a book-loving high school student, or a recent college graduate. Having said that, owing to the broad range of the topics of contention elucidated and the personal storytelling approach adopted, I believe this book can also be an intriguing reading for the general public who is keen on the global climate challenges and policy debates.

Finally, I would like to express my gratitude to not a few climate scientists and economists, referred to in this book and with many of whom I had personal encounters, who have made an enduring contribution to the humanity's knowledge on global climate challenges. I would also like to thank anonymous reviewers who read and provided constructive comments at any stage of the development of this book. The book would not have been a reality without the outstanding works by the editorial team at Palgrave Macmillan, including Wyndham H. Pain and Lavanya Devgun.

Bangkok, Thailand   S. Niggol Seo, Ph.D.

CONTENTS

About the Author

Professor S. Niggol Seo is a Natural Resource Economist who specializes in the study of global warming and globally shared goods. He received a Ph.D. degree in Environmental and Natural Resource Economics from Yale University in May 2006 with a dissertation on microbehavioral models of global warming. He held Professor positions in the UK, Spain, and Australia from 2006 to 2015. He has published over 120 (peer reviewed) articles on global warming economics, which includes seven books. He is currently at the Muaebak Institute of Global Warming Studies.

LIST OF FIGURES

LIST OF TABLES

Climate Change and Economics with Young Activists: An Introduction

1.1 THE EMERGENCE OF GENERATION-Z

When it comes to global warming, be it science, economics, or politics of, generation-Z has especially keen eyes. They were born at the threshold that lies between the twentieth century and the twenty-first century with all the promises and aspirations of the global community for the new millennium bestowed upon them. In particular, they came to this world after the world community had signed the first, and effectively the last, global warming treaty, namely the Kyoto Protocol, in December 1997 and made a heroic decision in July 2001 to implement the treaty across the Planet even without the US participation (UNFCCC 1997; Nordhaus 2001).

At the time of this writing, generation-Zers (gen-Zers henceforth) are either enrolled in colleges or in preparation for enrolling in colleges. They are the young generation of the global society *versus* the old generations born during the twentieth century. Every indication is that they are slowly emerging as a distinct as well as an assertive voice in the global community on many high-priority policy issues of today, global warming in particular.

Reflecting upon the events that have occurred in the world during the aforementioned time period, readers of this book may not find it difficult to conjecture what I am telling about here. Let me be more specific, just slightly though. Gen-Zers have grown up with the scientific debates on global warming, national politics, international negotiations, and civil

S. N. Seo, *Climate Change and Economics*,
https://doi.org/10.1007/978-3-030-66680-4_1

climate protests, all of which have, I am certain, helped sharpen their eyes, instincts, and intellectual abilities to dissect the problems. They witnessed through their fresh eyes, to name just some of the landmark events, the adoption of the Kyoto-Bonn Accord in July 2001, the Nobel Peace Prize awarded to Al Gore and the Intergovernmental Panel on Climate Change (IPCC) in 2007, the Kyoto Protocol's first phase implementation in January 2008, the shambles at Copenhagen Conference in December 2009, floating protesters at Cancun Conference in 2010, and the Paris Agreement in December 2015 amid the terrorist attack that rocked the city and the Planet (UNFCCC 2009, 2010, 2011, 2015; Seo 2017).

The young generation grew up experiencing the intense emotions associated with climate or natural catastrophes, including the Indian Ocean tsunami in 2004, the devastation of Hurricane Katrina in the US in 2005, the Fukushima-Daiichi nuclear disaster in Japan in 2011, and the bush fires in California and Australia in 2019 which seemed to rage ceaselessly. Each of these events ended up killing several thousands of people to several hundred thousands of people, in addition to millions of animals lost and countless houses destroyed.

As part of their education, unlike the older generations, gen-Zers learned as one of the first things in their life, to most of them I believe, at the grade of pre-schooling the following basic science: a historical global warming trend chart, a historical carbon dioxide concentration trajectory, a carbon cycle diagram, and greenhouse effects (NCEI 2019; Keeling et al. 2009). They also witnessed the award of the Nobel Prize to Al Gore and the aforementioned international group of climate scientists in 2007, to climate economist William Nordhaus in 2018, as well as the confident embrace of climate sciences and predictions by President Obama who held the office from January 2009 to January 2017 when they were teenagers.

Considering the milieu against which the new generation has come of age, it is not surprising that the Z-generation thinkers are equipped with strong opinions, one way or another, on the Planet's climate change and the policy negotiations to address it. The present author suspects that the gen-Zers have different perceptions and perspectives on the Planet's climate questions from other older generations including the millennials who were born during the twentieth century and before the Kyoto Protocol in 1997.

The signs of their perceptive presence that are emerging are plentiful. So, it would not be hard for you to encounter one if you really cared to see one. The most visible of them is perhaps the School Strike for the Climate, Skolstrejk för Klimatet in Swedish, the movement that held four world-wide strikes during the year 2019: March, May, and September. For the September strike, it is reported that about 4,500 strikes took place concurrently in 150 countries by about 4 to 6 million protesters (*The Guardian* 2019).

According to the media reports, many participants were school children enrolled in high schools, middle schools, and even elementary schools. Also, adult climate activists including college students widely participated in the global strikes. For the School Strike for the Climate which was originated from a Swedish girl who was 16 years old at the time, school students are urged to skip their schools every Friday to participate in the climate strike whose only goal is to save Earth from a Planetary climate crisis.

Young activists at the 2019 September strike proclaimed their concerns in stark terms which were equally enthusiastically carried by the world media. "We are the future and we deserve better," said a young activist from Thailand. "We are out here to reclaim our right to live, our right to breathe, and our right to exist," told a protester from India. Concurrently, the leader of the School Strike rebuked the audience at the United Nations with "You've stole my dreams and my childhood with your empty words."

Of all their proclamations, the sharpest thorn to the heart of any grown-up in the world who has a concern, however tiny and fickle it may be, on the global climate challenges may have been the single word, "future." Any adult including the present author would be flooded by a potpourri of emotions especially if the word were put forth by "a future generation" herself. When the strong emotions of the heart would settle down, many critical questions would likely await her or him, which have to be addressed intellectually, say, not by the heart.

Aptly titled "Climate Change and Economics: Engaging with Future Generations with Action Plans," this book is about these emotions of the heart on both sides, that is, the old generation and the young generation, as well as many intellectual questions that spring from the interactions, physical or mental, with the future generations, that is, generation-Z and other generations that will come after them.

Unlike many other books on climate change found in your bookshelves and libraries, I take a story-telling approach throughout this book, which means primarily that this book will not be written by a Planetary mind but by an individual citizen's mind who experiences all the things to be mentioned in this book, say, changes in the global climate, ecologies, personal economies, and livelihoods. I imagine that the present author is taking you to a stroll in numerous places across the Planet from as low as a mud-house in the hot tropics to as high as a granite city in the freezing northern country.

The final destination of the many strolls is a big-picture portrayal of climate change, economy, and livelihoods on the Planet. This big picture should, if it should be considered a success, be plain and at the same time persuasive to both the young and the old. It will be as easy as the van Goh's Sunflower and as intriguing as the Picasso's Guernica. In it, through the strolls, you will come across a bewildering rainforest, a river that never ends, an awe-inspiring water buffalo, the eye of a category-V deadly hurricane, a solar mirror way high up in the sky, an exceedingly remote island in the middle of the Pacific Ocean, a humble greenhouse in the Food Valley, everyone's favorite coral cays, among other things that would make you speechless.

1.2   Generation-Z and Future Generations

Let me begin first by clarifying the two essential terms on generational cohorts which will appear throughout this book: generation-Z and future generations. The generation-Z refers to the cohort of people born after around the dawn of the twenty-first century. A person that belongs to the generation-Z cohort is called a gen-Zer (Dimock 2019).

The generational cohort before generation-Z is referred to as the millennial generation. The millennials are those who were born during the period from the beginning of the 1980s to the middle of the 1990s. As such, nearly all millennials will remember the dawn of the new millennium as a memorable event during their primary school years. The millennials will be in their 30s or 40s, an adult, when they pick up this book by some serendipity.

The generational cohort before the millennials is called most often generation-X who were born between 1965 and 1980. The generational cohort before the generation-X is called baby boomers as they were born after the two world wars of the twentieth century. At the present time, the generation-X and the baby boomers are the old generations. I realize that the present author belongs to the generation-X.

As I described it before, the twenty-first-century threshold is marked by the signing of the Kyoto Protocol in December 1997 and the Bonn decision in July 2001 to implement the Protocol even with the US drop-out. As such, it is a salient landmark in the timeline of global climate challenges. The generation-Z is sometimes called iGeneration owing to the Internet and communication technology revolution since the mid-1990s such as personal computers (Microsoft), smartphones (Apple), social networking (Facebook), and artificial intelligence (Google).

Is there another generation after generation-Z and, if there is, what is it called? Surely, there will be another generation of people who are younger than gen-Zers, that is to say, as long as there is no humanity-ending catastrophe caused by climate change or something else. To them, gen-Zers will be greeted as an old generation. It appears that multiple terms are emerging to refer to this next generation, but we still need to wait and see how this next generation will be formed and interpreted.

The future generation is relatively defined. For the present generation at the time of this writing, the future generation is generation-Z. To generation-Z, the future generation will be the next generation after them, say, their own generation-Z or generation-Z^2. To generation-Z^2, there will be another future generation, say, generation-Z^3. The same can be said of generation-Z^3 and all the generations that follow.

For the sake of clarity of our conversations throughout this book, let's refer to all the generation cohorts beginning from generation-Z as "future generations" in a plural form. By contrast, "the future generation" in a singular form refers only to generation-Z.

In the above, I pointed to the unique experiences of gen-Zers with regard to the Planet's climate challenges. The present author imagines that gen-Zers will be the primary audience and readers of this book, consequently, all the explanations and presentations will be put forth at their eye level.

Having said that, a more proper timeframe for an analysis of climate change and economics is that of future generations, that is, encompassing of all generations from generation-Z onwards. Readers will soon find out that the analyses in this book are conducted in most cases, with some exceptions, in the timeframe of the future generations. To give you just a single example, an analysis of the inter-generational burden sharing on the Planet's climate challenges provided in Chapter 13 covers the future generations that would come to live in the next 300 years from today.

1.3   Climate Change Standing Alone

For gen-Zers, climate change, the term and the science of it, has always been around there with them. This is not the case for the other generations. The science of climate change was not there when the preceding generations were born and grew up. For the older generations, the science of climate change came to be a publicly discussed science knowledge by the time they hit the age of a high teen, or a college student, or even at a later stage in their life. Owing to this, generation-Z will certainly feel more affinity with the science of climate change than the preceding generations.

Notwithstanding, that something has always been there with generation-Z without them having the opportunity to wait for it or think critically about it before accepting the truth of it does not always signal a positive consequence. Cognitively, to conjecture one, it is possible that a gen-Zer has developed an innate bias, one way or another, toward the science of climate change. In other words, she/he may have a subconscious prejudice about the science of climate change.

From another viewpoint, the science of climate change has always been taught to them as a global crisis that endangers the future of humanity and the Planet. As children, gen-Zers may have developed the equivalence between the future of humanity and their future, mistakenly concluding that they are indeed "the" future generation implied in the science.

The equivalence, more often than not implied in the climate science reports, may have engendered the sense of victimhood among gen-Zers. That is, they may have reasoned that climate change or global warming will harm or even destroy Planet Earth in the future to which they belong and further which is their home. The blame, they may have reasoned further, belongs to the adults, say, the older generations.

This is a general trait in human intelligence or mental capacity. Until you have the sufficient time to think critically about something, you will always be in possession of one prejudice or another nearly innately, in other words, without your own noticing of it. So, it is quite pertinent and meaningful that gen-Zers take the opportunity to reexamine their knowledge on the phenomenon of climate change, through a college education or a self-study, even if it would risk your social networking influence. I am sure that you, as a gen-Zer, will find many things and aspects that are novel to you or even contrary to the ways that you have known them before.

The science of climate change is complex, to say the least, although it is not the only complex thing in your life. This is because the science has to address the entire Planet whose parts and elements are so myriad and disparate in many ways while their inter-relationships are complicated. Of the relationships, the most difficult one involves the relationships between the natural system and the anthropogenic systems. To make things more complicated, it may and would involve predicting the relationships between Planet Earth and the factors outside the Planet, for example, solar radiation, particles in the universe, and cloud formations.

As an introductory chapter of this book, I will, from now on, present you the science of climate change in a manner as succinct and lucid as possible. This will be followed by the description of the economics and then of the livelihood aspects of climate change in a similarly concise manner. Think of these as a presentation at the basecamp of Mt. Everest before your hike!

Let's begin with the main character of climate change science: carbon dioxide (CO_2). You may all know already that the chemical compound is responsible for global warming and is released by the power-plants that burn fossil fuels for energy and electricity generations. You may also know that oil, natural gas, and coals are those fossil fuels.

You may not know, at least some of you, that carbon dioxide is everywhere on Earth. Carbon is the second most abundant element in the human body next to oxygen. Both carbon and oxygen account for roughly 84% of the mass of a human body. All plants on Earth, that is, trees, shrubs, grasses, and flowers, store carbon in their plant bodies (To be explained in detail in Chapter 3 of this book). The amount of carbon stored in the soils of Earth is "massive." The oceans of the Planet store carbon at ocean floors and lower oceans in vast amounts (To be explained in detail in Chapter 5 of this book) (Ciais et al. 2013).

From another angle, carbon, combined with oxygen in the air, provides essential energy to the Planet and humanity. Specifically, through carbon dioxide, plants obtain energy needed for life via a photosynthesis mechanism. In turn, humans as well as animals absorb carbohydrates by consuming plant materials to acquire necessary energy for their survival and actions.

The point is that carbon dioxide, the central character in the literature of climate change, possesses many faces when it relates to a shift in the global climate system and its consequences. This is the case not just for carbon dioxide, but for many key variables in climate science, to mention just some of them, measuring the Planet's temperature, measuring green-house gas concentrations in the Earth atmosphere, predicting a future climate system, predicting hurricane occurrences and other disasters in the future, estimating a mass extinction of species, and predicting changes in food supply.

The science of climate change is at heart the science of greenhouse effect. A group of chemicals, referred to as greenhouse gases (GHG) collectively, forms a greenhouse-like blanket in the Earth's lower atmosphere which has the consequence of warming the Planet. The greenhouse effect, that is, the Planet-warming effect, occurs because the greenhouse-like layer traps the infrared (longwave) solar radiation reflected by the Earth's surface although it is transparent enough to pass sunlight (shortwave radiation) through to reach the Earth surface (IPCC 1990; UNFCCC 1992).

The cohort of greenhouse gases is carbon dioxide (CO_2), methane (CH_4), nitrous oxides (N_2O), water vapor (H_2O), and certain fluorinated gases such as hydrofluorocarbons (HFCs). So, what causes climate change is not just carbon dioxide. Many other activities, natural and anthropogenic, that do not lead to carbon dioxide emissions are also to blame for the Planet-warming effect. To give you a salient example, farm animals such as cattle, goats, and sheep release methane when they burp or fart. When you consume steak or mutton in a nice corner diner, you are responsible for some Planet-warming, however small it may be. For another, cooling agents in a refrigerator and an air-conditioner such as hydrofluorocarbons (HFCs) have Earth-heating effects. If you are reading this book in a well air-conditioned library of your university, you may be accused, by your activist friend, of causing some global warming! The point that I want to get across to you is many things in climate change are personal.

After all, how much the Earth's climate has changed? In Fig. 1.1, the present author draws two trajectories of the Earth's average temperature. One is the instrumental temperature data based on the records from the ground weather stations dispersed across the Earth surface and the other is the satellite temperature data based on the records from Earth-orbiting satellites. In the figure, the ground weather station data are from the National Aeronautics and Space Administration (NASA) of the US and the satellite data are from the global temperature report by the team at the University of Alabama at Huntsville also funded by the NASA (Schmidt et al. 2016; UAH 2020).

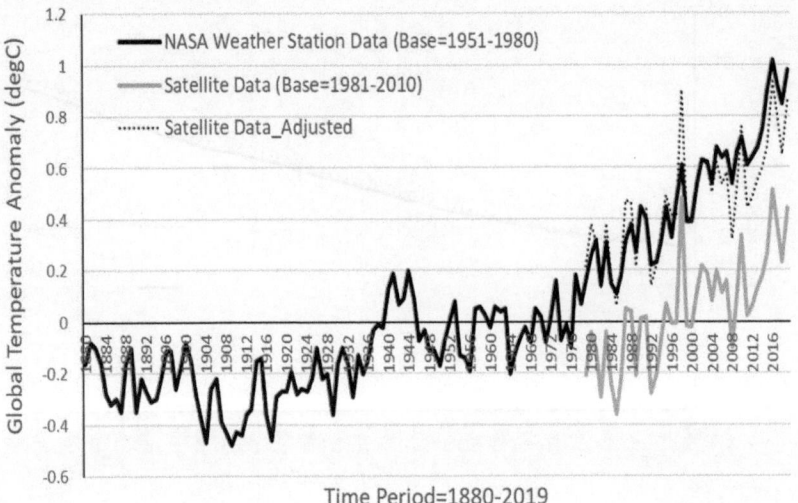

Fig. 1.1 The Earth's temperature changes since 1880: weather station versus satellite (*Note* Data made available to the public and researchers by the Carbon Dioxide Information Analysis Center [CDIAC] of the US Government)

The ground weather station data cover the entire record period from 1880 to 2019, for which the baseline period is 1951–1980. The data are expressed as anomalies or deviations from the average of the baseline period. The satellite data cover the entire record period from 1978 to 2019, for which the baseline period is 1981 to 2010. Since the two data rely on different baseline periods, it is not easy to compare them directly. So, the present author lifted up the satellite data by the difference between the average of the weather station and that of the satellite data during the 1981–2010 period.

The two temperature records are by and large in agreement as far as a gradually increasing global temperature trend is concerned. The trajectories reveal that the global average temperature has increased by about 0.8 °C in the ground station data and by about 0.4 °C in the satellite data from their aforementioned respective baselines.

Having said that, it is also noticeable that there are notable discrepancies in some peak and trough years: concretely, the peak year in 1998, the trough years in 2008 and 2010. The peak year temperature in 1998

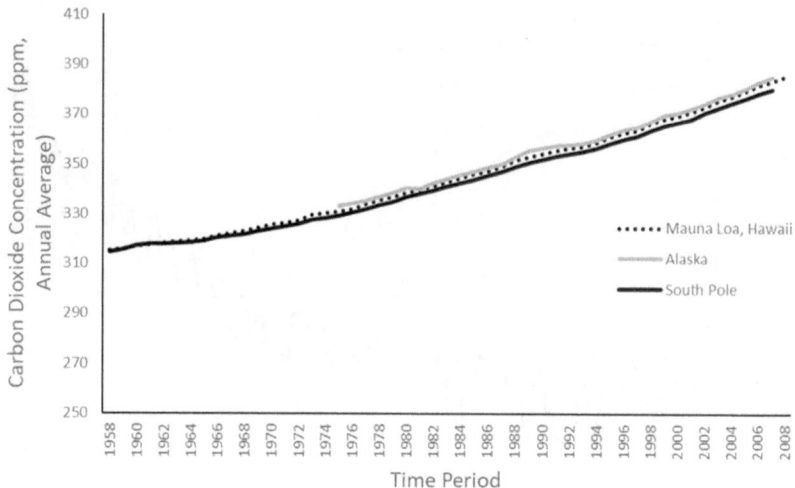

Fig. 1.2 CO$_2$ records from multiple sites (*Note* Data made available to the public and researchers by the Carbon Dioxide Information Analysis Center [CDIAC] of the US Government)

is far higher in the satellite record and the trough year temperatures are noticeably lower in the satellite record in 2008 and 2010.

What has caused the rises in the global average temperature shown in Fig. 1.1? This is where the main character comes in: carbon dioxide. The long-term trajectory of the carbon dioxide concentration in the global atmosphere is shown in Fig. 1.2, which is the highly acclaimed Keeling Curve. Charles Keeling has measured continuously CO$_2$ concentration from the middle of the 1950s at the Mauna Loa Observatory in Hawaii (Keeling et al. 2009). The reading during the initial years from the instrument he built was about 315 ppm (parts per million), which has continuously climbed to about 380 ppm in 2008, and further to about 410 ppm in 2020. The other measurements from the Alaska and the South Pole observatories show mostly identical trends.

1.4   CLIMATE CHANGE IN TANDEM WITH ECONOMY

The science of climate change was inextricably tied via the Keeling Curve to the economy, more explicitly, the economic growth whose salient

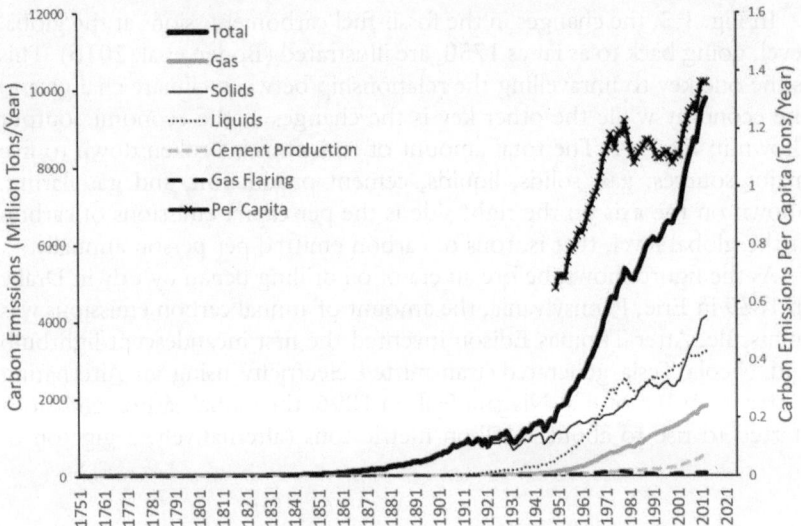

Fig. 1.3 Annual global fossil-fuel carbon emissions since 1750 (*Note* Data made available to the public and researchers by the Carbon Dioxide Information Analysis Center [CDIAC] of the US Government)

byproduct is carbon dioxide. The carbon dioxide concentration had been stable for thousands of years on Earth at roughly 280 ppm. When the era of rapid economic growth kicked off at the end of the nineteenth century with new inventions of a crude oil extraction, electricity, and automobiles, the CO_2 concentration in the global atmosphere kick-started its rapid upward trend.

The remarkable economic growth during the twentieth century was spurred by the energy generated from fossil fuels. The major inventions of crude oil extractions, powerplants with a nationwide electrical grid, internal combustion engines with a public road network, airplanes, and radio communications have made possible the economic success of the twentieth century. As an unforeseen consequence, however, the amount of carbon dioxide emitted into the atmosphere from burning of fossil fuels has increased sharply over this time period. By the early 1970s, pioneering economists started to have a glimpse of the carbon dioxide problem of the Planet caused by the economic growth (Nordhaus 1977).

In Fig. 1.3, the changes in the fossil-fuel carbon emissions at the global level, going back to as far as 1750, are illustrated (Boden et al. 2016). This is the one key to unravelling the relationship between climate change and the economy while the other key is the changes in the economic output shown in Fig. 1.4. The total amount of emissions is broken down to five major sources: gas, solids, liquids, cement production, and gas flaring. Shown on the axis on the right side is the per capita emissions of carbon at the global level, that is, tons of carbon emitted per person annually.

As the figure shows, before an era of oil drilling began by Edwin Drake in 1859 in Erie, Pennsylvania, the amount of annual carbon emissions was miniscule. After Thomas Edison invented the first incandescent lightbulb and Nicola Tesla generated/transmitted electricity using an Alternating Current (AC) motor at Niagara Falls in 1896, the global carbon emissions started to rise to about 1 billion metric tons (alternatively, a gigaton of

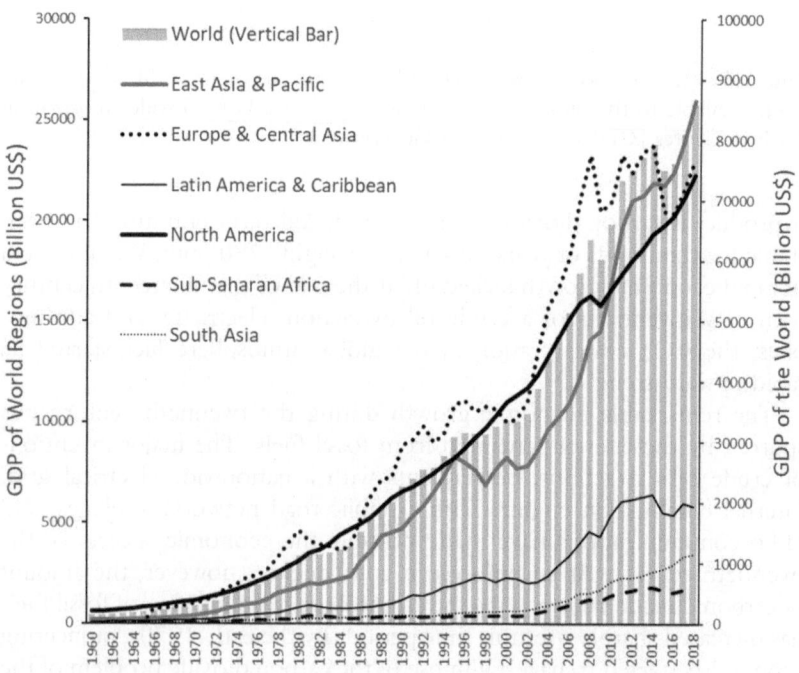

Fig. 1.4 World GDP and Regional GDPs since 1960

carbon [GtC]) by the dawn of the twentieth century. After Ford Motor Company produced the first mass-produced internal combustion engine, Model T, during the 1910s and 1920s, the global carbon emissions can be seen in the figure to rise at a faster rate. After the two world wars, the amount of annual global carbon emissions can be seen to rise at even a faster rate. By 2020, it stands at roughly 10 GtC/year, which is about 10 times the amount at the beginning of the twentieth century.

Underneath the steep rise in the global carbon emissions from burning fossil fuels especially during the 2nd half of the twentieth century lies an increase in the economic production globally. In Fig. 1.4, the world total new production annually, more formally the Gross Domestic Product (GDP), since 1960 is drawn in vertical gray bars, along with the regional GDPs of the world major regions overlaid as line trajectories (World Bank 2020). You may refer to the Appendix of this book if you need a clarification of the concept of the GDP and other economic output measures such as the Green GDP.

The figure shows that the world GDP has grown almost exponentially from US$1 trillion in 1960 to US$75 trillion in 2016, a 7500% expansion over this period (World Bank 2020). Compare this economic expansion with the carbon dioxide increase in Fig. 1.3 which increased by about 1000% since the beginning of the twentieth century! The regional GDP trajectories tell us further that the economies in East Asia & Pacific, North America, and Europe and Central Asia have expanded vastly over this time period while the economies in South Asia are just getting started to catch up with the three high-income regions. The economies in Latin America and Caribbean as well as Sub-Saharan Africa appear to exhibit the weakest momentum for economic growth.

You may ask why the fossil fuels have been so pivotal for the economic growth and, say, the economic well-being of the people on Earth, considering the strong correlation between the carbon emissions trajectory in Fig. 1.3 and the global economic output trajectory in Fig. 1.4. The answer lies in the amount of energy that can be generated from burning fossil fuels, in terms of both the total energy generated and the high energy density. The potential energy is so large as to make it possible for humanity to have electricity supplied to all houses and factories, to run all automobiles and planes for many centuries, and to provide sufficient heat to make steels, cements, and bricks, all for many centuries. The amount of energy is measured by Joule (J) and British Thermal Unit (BTU), which we will have an opportunity to review with formal concepts and real-world statistics at the energy-focus chapters of the book, specifically, Chapters 7 and 8 (USEIA 2020).

As per the high energy density of the fossil fuels, consider this: It would take only a gallon of gasoline to move a 2000-kg-heavy vehicle for the distance of 30 km. It would take only a coffee break to complete the job. By contrast, such a job may be completed by a dozen strong donkeys for many days, if not months.

In addition to the causal relationship between the global economy and the climate change implied in Figs. 1.3 and 1.4, there is another essential causal relationship between the two, which is often overlooked by both observers and researchers. Specifically, the carbon dioxide, released into the atmosphere, is a fodder to the photosynthesis of plants—trees, shrubs, and grasses—through which the chemical energy, which is sugars, needed for their survival is manufactured by converting the Sun's light energy (Schlesinger 1997). An increase in the amount of carbon dioxide in the atmosphere causes an enhancement of the photosynthesis in the plant biomes, thereby making them healthier and bigger.

The impact of the carbon dioxide's enhancement of photosynthesis is not limited to plants, algae, and bacteria. Since animals as well as humans consume plants, grasses, and tree products through which they acquire energy as well as other micronutrients needed for their survival, an enhanced photosynthesis of the plant biomes will end up enhancing the well-being of humans and other animals on the Planet. More directly speaking, the economic well-being of the Planet will be improved, *ceteris paribus*, because of a large atmospheric increase in the amount of carbon dioxide.

How large are the impacts of the atmospheric carbon dioxide increase on the plant ecosystems? The impacts of a doubling of carbon dioxide concentration on the ecosystems are studied through either a laboratory-based experiment or an open-air field experiment. The latter is often called the FACE experiment short for the Free Air CO_2 Enrichment experiment. The results from a meta-analysis of the FACE experiments during the 15-year period examined by the authors are summarized in Fig. 1.5 (Ainsworth and Long 2004). The impacts are measured, as shown in the figure, by a suite of ecological indicators such as Leaf Area Index (LAI), Dry Matter Production (DMP), Stem Diameter (SD), Branch Number (BN), Leaf Number (LN), Yield (YD), and Height (HT). The black bars in the figure marked as "Plants, General" are the summary results for

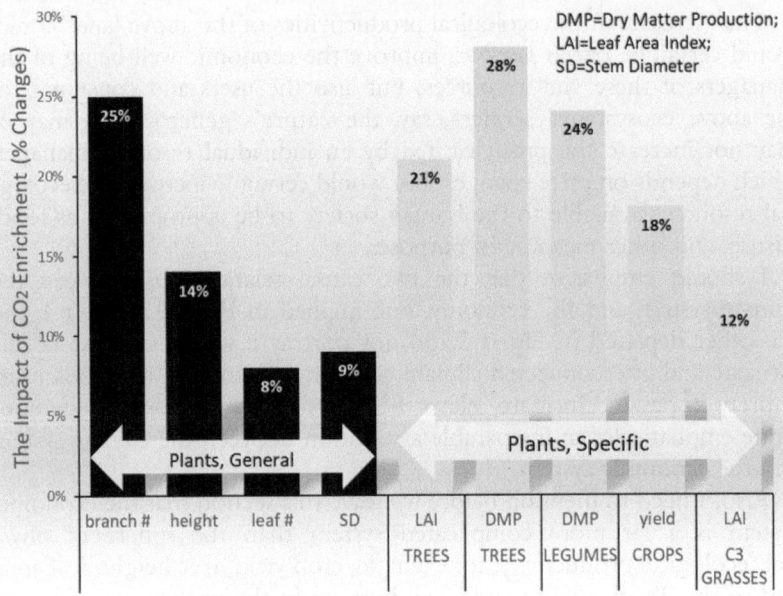

Fig. 1.5 Impacts of climate change on ecosystems

all plant biomes while the gray bars marked as "Plants, Specific" are the summary results for specific plant biomes.

The figure reveals that the carbon dioxide doubling in concentration is beneficial to the plant ecosystems. For trees, there would be a 21% increase in the leaf area index as well as a 28% increase in the dry matter production, both of which are attributable to the doubling of carbon dioxide concentration. For crops, there would be a 20% increase in crop yields averaged across all crops experimented by the FACE researchers. For grasses, the LAI would be increased by 15% while the DMP for legumes would be increased by 23%.

At the level of the plants in general, the meta-analysis reveals, as shown in the figure as black bars, that the plant height is observed to increase by 15% on average, the stem diameter by 9%, the branch number by 26%, and the leaf number by 8%. These beneficial changes in the four ecological productivity indicators for the plant biomes are attributable to the doubling of carbon dioxide concentration created at respective experimental sites by the FACE researchers.

The increases in the ecological productivities of the above land biomes would certainly, *ceteris paribus*, improve the economic well-being of the managers of these vital resources, but also the users and consumers of the above ecosystems' services, say, the nature's generosity. It may or may not increase the profit earned by an individual resource manager, which depends on price changes, but would certainly increase the ecological resources available to the human society to be appropriated as food, leisure, and other meaningful purposes.

I should emphasize that the two causal relationships between the climate system and the economy, one implied in Figs. 1.3 and 1.4 and the other depicted in Fig. 1.5, do not portray a whole universe of the intricate and interconnected climate-economy system that this book in its entirety hopes to elucidate. Nevertheless, the two relationships demonstrate emphatically an inseparable association between the climate system and the economic system.

Also, I need to mention before we leave this section that the economic system is a far more complicated system than the sphere of physical/ecological productivity, for example, crop yield, tree height, leaf area, and so on. To give you an idea, an increase in the production of a corn crop by a farmer in Thailand may turn out to be a harmful thing to the farmer because of a price decline caused by a large oversupply of corn to the market. The farm profit will fall despite the increased harvest and sale. The point here is that the economic system can only be correctly analyzed with a careful modeling of prices and the market.

1.5   Climate Change and Life

The descriptions up to this point may not reverberate in your heart, at least not yet. You may ask: Does the change in the world's GDP have anything to do with my own life?; I am not a farmer, nor a natural resource manager. Why should I care about the changes in the crop yields or the number of tree leaves and branches?; How does the increase in carbon dioxide concentration in the global atmosphere affect my ways of life?; Does the increase in long-term global average temperature harm my ways of life at around my neighborhood?; Will global warming make my daily lives and of animals intolerable and vulnerable? There will arise, which is quite natural, a long list of questions in your mind.

Many, if not most, of the questions will pertain to personal economics as well as non-market economics, which I need to clarify. The former is about the economic consequences of climate change at the personal level in contrast to those at the economy-level. The latter is about the economic consequences of climate change on the activities and goods not traded in the marketplaces which nonetheless would impose changes in monetary values of your assets. As long as you do not become clear on these two aspects of climate change, you will not be able to truly appreciate the debates and contentions on climate change.

From the perspective of the personal economics, the problem of climate change is often described to be a pervasive economic issue. Pervasiveness here refers to the fact that nearly all of your economic activities have some consequences of carbon dioxide and other greenhouse gas emissions as well as the fact that nearly all of your activities at the personal level would be impacted by changes in the climate system.

Let me be more specific. When you enter your house, you will turn on the lights, through which you consume electricity generated from power-plants by burning coal or natural gas. This releases carbon dioxide. When you drive your automobile, it will burn gasoline or diesel to create necessary energy to move the car, through which carbon dioxide is released. When you eat a beefsteak and a corn salad, the productions of the source products such as beef cow and corns by the farmers will release greenhouse gases such as carbon dioxide and methane. When you purchase your flannel shirts or a Cashmere suit, the productions of source materials will release greenhouse gases into the atmosphere. Your reliance on a mobile phone cannot be sustained without charging the battery every night, for which you need electricity generated from burning most likely fossil fuels. While you read this book, you notice that the papers used for the book making are made of the pulp which is produced from cutting trees, which releases carbon dioxide.

You may insist that the battery in your mobile phone as well as the electricity in your home is provided by clean energy sources, say, solar panels on the rooftops. Or you may argue that you always ride a bike to your school instead of driving your own automobile. Such behavioral changes could reduce the amount of your emissions of carbon dioxide, but not eliminate them. The solar panels should be made from many source materials, installed across a vast land area, often cause a blackout, and sometimes cause a landslide, all of which make solar energy not as carbon dioxide free as we may hope for.

At a subtler level, an adoption of a low-carbon energy or technology may force your other behavioral changes which may end up increasing your carbon dioxide emissions. An adoption of a bicycle ride to your school, to give you an illuminating example, may increase your consumption of water and other drinks or make you to take a shower more frequently, all of which tend to increase your consumption of electricity and water.

The pervasiveness of climate change as an economic issue may be most strongly felt by the farmers and natural resource managers especially in the developing countries in Africa, Latin America, and South Asia. Their incomes and livelihoods depend to a large extent on outdoor conditions, the most important of which are climate/weather and soils (World Bank 2020). A farmer in Sub-Saharan Africa has a large pool of options with some of which she/he can manage her farm: scores of crops, dozens of farm animals, and a multiplicity of trees and tree products. A choice of one portfolio by the farmer against the other portfolios would have important consequences on the amount of carbon dioxide emissions she/he emits as well as the profit she/he can earn from the farm (Seo 2016a, b, c).

Because of the immediate dollar consequences, farmers and natural resource managers will be conscious in their choices to choose the best portfolio taking into considerations the climate regime in their farms. To continue to explain their decisions concretely, let's say that a farmer in Sub-Saharan Africa has three portfolios to choose from. Let's assume that portfolio I earns 1,000 dollars per year annually, portfolio II earns 1,500 dollars per year annually, and portfolio III earns 2,000 dollars per year annually, all minus the expenses, given the current climate regime at the farm. She/he would choose portfolio III, with other conditions being equal.

In Fig. 1.6, I present such deliberate choices of the Sub-Saharan African farmers under two different climate change scenarios (Seo 2012). There are in the figure three types of agricultural systems. System I is a crops-only portfolio which specializes in crops. System II is a mixed crops-livestock portfolio which diversifies into both crops and farm animals. System III is a livestock-only portfolio which specializes in farm animals.

The two scenarios of climate change examined are, on the left panel, an increase in rainfall variability by 30% measured by the coefficient of variation in yearly precipitation (CVP) and, on the right panel, an increase in diurnal temperature range (DTR) by 3 degrees in Celsius. Both climate variables are constructed from the 30-year data of the corresponding weather variables. The CVP is the degree of variability of monthly precipitation for the 30-year period and the DTR is the difference between

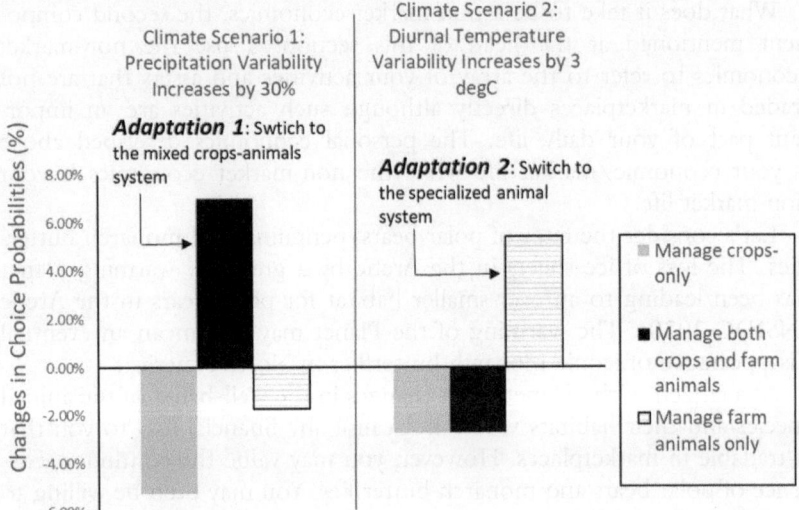

Fig. 1.6 Personal decisions to adapt to an increased climate risk in Sub-Saharan Africa

the daily maximum and the daily minimum temperature averaged over 30-year period.

The figure illustrates how Sub-Saharan farmers have adapted to these changes in the climate regime and will do so again faced with the same risk situations. When the rainfall risk captured by the CVP increase by 30%, farmers reduce the specialized portfolios: the crops-only and the livestock-only. Instead, they increase the diversified portfolio: the mixed crops-livestock by 7% points, which is marked as Adaptation 1 in the figure. This decision seems to indicate that the diversified system suffers less than the specialized systems from a high rainfall variability.

The adaptative behavioral changes by the farmers to the increase in the DTR, that is, the temperature risk, are shown to be different. Sub-Saharan farmers reduce both the crops-only portfolio and the crops-livestock portfolio. Instead, they increase the livestock-only portfolio by 6% points, which is marked as Adaptation 2 in the figure. This decision seems to indicate that farm animals are more resilient to daily temperature variability than crops, that is, grains, beans, and vegetables.

What does it take to be a non-market economics, the second component mentioned at the head of this section? I use the non-market economics to refer to the array of your activities and assets that are not traded in marketplaces directly although such activities are an important part of your daily life. The personal economics described above is your economic/market life while the non-market economics is your non-market life.

Let's consider the fates of polar bears, penguins, and monarch butterflies. The loss of ice sheets in the Arctic by a gradually warming Planet has been leading to an ever-smaller habitat for polar bears in the Arctic (NSIDC 2020). The warming of the Planet may also mean an eventual disappearance of iconic monarch butterflies in North America.

As a citizen of the Planet, such changes in the well-being of the animal species and their habitats would not cause any financial loss to you that is tradable in marketplaces. However, you may value the continued existence of polar bears and monarch butterflies. You may even be willing to contribute financially to a conservation agency for preservation of these magnificent species. Your donation would not be in any way related to the market value, that is, sale price of a polar bear or a monarch butterfly.

The question of biodiversity and mass extinction will be taken up for an extensive discussion in Chapters 9 and 10. I will review the literature on habitat losses as well as biodiversity losses, which will shed light on a paradox of biodiversity loss or mass extinction in a changing climate. The second chapter of the two will be centered on marine species and commercial fisheries including coral reefs. The chapters would also provide an answer to the question of whether a drastic and costly policy intervention at the global scale is needed to save endangered species via a "global deal" for nature (Dinerstein et al. 2019; CBD 2020).

1.6   What Lies Ahead in the Book

This book titled "Climate Change and Economics: Engaging with Future Generations with Action Plans" is written as an introductory textbook to gen-Zers and the future generations on climate change, economy, and life. I gave a succinct exposition of the core of each of the three components in Sect. 1.3 (climate change), Sect. 1.4 (economy), and Sect. 1.5 (livelihoods).

As I mentioned before, I take a personal story-telling approach in writing this book. I strived to base the contents of this book on the personal experiences of mine as well as the others' expressed in the literature. Further, this means that the book is a big mosaic of fourteen distinct stories, each of which is told in each chapter. To be specific, the following stories are told in this book:

> Chapter 1: Climate Change and Economics with Young Activists: An Introduction;
> Chapter 2: Friends of Animals: A Story of Sheep and Goats in the Sahel in Sub-Saharan Africa;
> Chapter 3: Giving Forests: A Tale of Amazon Rainforests and Congo River Forests;
> Chapter 4: Indian Monsoon: A Tale of Indian Water Buffaloes, Goats, and High-yield Rice;
> Chapter 5: A Refuge from Oceans and Hurricanes: A Story of Cyclone Shelters in Bangladesh Abutting the Bay of Bengal;
> Chapter 6: Sublime Grasslands: A Story of the Pampas, Prairie, Steppe, and Savannas where Animals Graze;
> Chapter 7: Energy Revolutions: A Story of the Three Gorges Dam in China;
> Chapter 8: Backstop Technologies: A Story of a Humble Greenhouse with Surprises;
> Chapter 9: A Story of Polar Bears and Penguins: A Paradox of Biodiversity and Climate Change;
> Chapter 10: A Story of Coral Reefs, Nemo, and Fisheries: On Biodiversity Loss and Mass Extinction;
> Chapter 11: A Story of Infectious Diseases and Pandemics: Will Climate Change Increase Deadly Viruses?;
> Chapter 12: Climate Negotiations: The Science of a Big Deal?
> Chapter 13: Climate Change and Economics with Young Enthusiasts: Inter-generational Gaps and Burden Sharing;
> Appendix: A Brief Exposition of the Essential Economic Theories Used in this Book.

The topics of the chapters are chosen, as you may be able to feel from the above list, in a way to be encompassing of all the major climate change issues, from both science and economics. In addition, the chapter topics are carefully selected to cover all regions of the Planet as well as valuable resources of the Planet, including the world geopolitical regions (Chapters 12 and 11), 8 world oceans (Chapters 5 and 10), two polar regions (Chapter 9), tropical rainforests (Chapter 3), major grasslands (Chapter 6), awe-inspiring rivers and mountains (Chapter 7), farm animals (Chapter 2), species of the Planet (Chapter 9), viruses and infectious diseases (Chapter 11), and technological innovations (Chapters 8 and 11).

Because I hope to present Planet Earth to you in a visual manner as well as what are occurring in it, the book is well illustrated with multiple geographic maps in many of the chapters. I wish you could have the feel of Planet Earth more tangibly after reading through this book and the book could give you a motivation to explore many things described in this book by yourself and to become a guardian of Earth.

Lastly, the book has an Appendix chapter to which you can refer while you are reading this book and find some terms unclear, titled "Appendix: A Brief Exposition of the Essential Economic Theories Used in this Book." I provide a brief explanation of the economic terms and theories used frequently in this book to help you read through the book more easily.

With this, I think I have said enough, for now. I wish you an amazing journey through this book of the Planet and every corner of it.

1.7   CHAPTER HIGHLIGHTS

- This chapter provides an introduction to and an overview of the book which has dual goals. One is to write a textbook of climate change and economics for young readers and the other is to address the inter-generational issues in the literature thereof.
- This chapter starts with the description of the unique experiences and recent activism of generation-Z, the description of the generational cohorts, and the definition of future generations.
- This is followed by a big-picture overview of climate sciences. Global average temperature trends are explained and the Keeling Curve of carbon dioxide concentration is described.
- The climate-economy connection is elaborated by, first, the economic growth as a cause of climate change and, second, the economic damage and benefit as a consequence of climate change.

- At the individual level, this chapter describes how individual citizens live with and adapt to the climate system which has never remained unchanging at any time in Earth history.
- The ways and extents with which such individual experiences as well as community experiences are a critical element for the world's responses to global climate challenges are a central message of this introductory chapter.

REFERENCES

Ainsworth, Elizabeth A., and Stephen P. Long. 2004. What Have We Learned from 15 Years of Free-Air CO_2 Enrichment (FACE)? A Meta-Analytic Review of the Responses of Photosynthesis, Canopy Properties and Plant Production to Rising CO_2. *New Phytologist* 165 (2): 351–72. https://doi.org/10.1111/j.1469-8137.2004.01224.x.

Boden, T.A., G. Marland, and R.J. Andres. 2016. *Global, Regional, and National Fossil-Fuel CO_2 Emissions*. Oak Ridge, TN: Carbon Dioxide Information Analysis Center, Oak Ridge National Laboratory. https://doi.org/10.3334/CDIAC/00001_V2016.

Ciais, P., C. Sabine, G. Bala, L. Bopp, V. Brovkin, J. Canadell, A. Chhabra, R. DeFries, J. Galloway, M. Heimann, C. Jones, C. Le Quéré, R.B. Myneni, S. Piao, and P. Thornton. 2013. Carbon and Other Biogeochemical Cycles. In *Climate Change 2013: The Physical Science Basis*. Cambridge: Cambridge University Press.

Convention on Biological Diversity (CBD). 2020. *Zero Draft of the Post-2020 Global Biodiversity Framework*. New York, NY: United Nations.

Dimock, Michael. 2019. *Defining Generations: Where Millennials End and Generation Z Begins*. Washington, DC: Pew Research Center.

Dinerstein, E., C. Vynne, E. Sala, A.R. Joshi, S. Fernando, T.E. Lovejoy, J. Mayorga, et al. 2019. A Global Deal for Nature: Guiding Principles, Milestones, and Targets. *Science Advances* 5 (4): eaaw2869. https://doi.org/10.1126/sciadv.aaw2869.

Intergovernmental Panel on Climate Change (IPCC). 1990. *Climate Change: The IPCC Scientific Assessment*. Cambridge: Cambridge University Press.

Keeling, R.F., S.C. Piper, A.F. Bollenbacher, and J.S. Walker. 2009. Atmospheric CO2 Records from Sites in the SIO Air Sampling Network. In *Trends: A Compendium of Data on Global Change*. Oak Ridge, TN: Carbon Dioxide Information Analysis Center, Oak Ridge National Laboratory. https://doi.org/10.3334/cdiac/atg.035.

National Centers for Environmental Information (NCEI). 2019. *State of the Climate: Global Climate Report-Annual 2019*. Washington DC: The National Oceanic and Atmospheric Administration.

National Snow and Ice Data Center (NSIDC). 2020. *Sea Ice Index: Arctic- and Antarctic-Wide Changes in Sea Ice*. Boulder, CO: NSIDC.

Nordhaus, William D. 1977. The Economic Growth and Climate: The Carbon Dioxide Problem. *American Economic Review* 67: 341–346.

Nordhaus, William D. 2001. Global Warming Economics. *Science* 294 (5545): 1283–84. https://doi.org/10.1126/science.1065007.

Schlesinger, William H. 1997. *Biogeochemistry: An Analysis of Global Change*, 2nd ed. San Diego, CA: Academic Press.

Schmidt, G., R. Ruedy, A. Persin, M. Sato, and K. Lo. 2016. NASA GISS Surface Temperature (GISTEMP) Analysis. In *Trends: A Compendium of Data on Global Change*. Oak Ridge, TN: Carbon Dioxide Information Analysis Center, Oak Ridge National Laboratory. https://doi.org/10.3334/cdiac/cli.001.

Seo, S. Niggol. 2012. Decision Making Under Climate Risks: An Analysis of Sub-Saharan Farmers' Adaptation Behaviors. *Weather, Climate, and Society* 4 (4): 285–99. https://doi.org/10.1175/wcas-d-12-00024.1.

Seo, S. Niggol. 2016a. *Microbehavioral Econometric Methods: Theories, Models, and Applications for the Study of Environmental and Natural Resources*. London: Academic Press.

Seo, S. Niggol. 2016b. Modeling Farmer Adaptations to Climate Change in South America: A Micro-Behavioral Economic Perspective. *Environmental and Ecological Statistics* 23 (1): 1–21. https://doi.org/10.1007/s10651-015-0320-0.

Seo, S. Niggol. 2016c. The Micro-Behavioral Framework for Estimating Total Damage of Global Warming on Natural Resource Enterprises with Full Adaptations. *Journal of Agricultural, Biological, and Environmental Statistics* 21 (2): 328–47. https://doi.org/10.1007/s13253-016-0249-2.

Seo, S. Niggol. 2017. Beyond the Paris Agreement: Climate Change Policy Negotiations and Future Directions. *Regional Science Policy & Practice* 9 (2): 121–40. https://doi.org/10.1111/rsp3.12090.

The Guardian. 2019. *Across the Globe, Millions Join biggest Climate Protest Ever*. London: The Guardian. Published on September 21, 2019.

United Nations Framework Convention on Climate Change (UNFCCC). 1992. *United Nations Framework Convention on Climate Change*. New York: UNFCCC.

United Nations Framework Convention on Climate Change (UNFCCC). 1997. *Kyoto Protocol to the United Nations Framework Convention on Climate Change*. New York: UNFCCC.

United Nations Framework Convention on Climate Change (UNFCCC). 2009. *Copenhagen Accord*. New York: UNFCCC.

United Nations Framework Convention on Climate Change (UNFCCC). 2010. *Cancun Agreements*. New York: UNFCCC.

United Nations Framework Convention on Climate Change (UNFCCC). 2011. *The Durban Platform for Enhanced Action*. New York: UNFCCC.

United Nations Framework Convention on Climate Change (UNFCCC). 2015. *The Paris Agreement. Conference of the Parties (COP) 21*. New York: UNFCCC.

United States Energy Information Administration (US EIA). 2020. *Energy Explained*. Washington, DC: US EIA.

University of Alabama at Huntsville (UAH). 2020. *Global Temperature Record*. Huntsville, AL: UAH. Available at: http://nsstc.uah.edu/climate/.

World Bank. 2020. *World Bank Development Indicators*. Washington, DC: The World Bank.

CHAPTER 2

Friends of Animals: A Story of Sheep and Goats in the Sahel in Sub-Saharan Africa

2.1 Sub-Sahara and the Sahel

I feel like I can start telling my stories on climate change and economics with a story of farmers in the Sahelian Africa who raise a host of farm animals in their farms, not to mention a large basket of grains, vegetables, trees, and forest products. It may not seem strange to you since Sub-Saharan Africa is often on the headlines of any climate change related news and reports. Above all, Africa is the hottest continent of all on the Planet while the Sahel and Central Africa is the hottest place on Earth where the Equator cuts through between the two regions.

The story of farm animals in the Sahelian region is fascinating in the milieu of the global warming research, which will be clarified throughout this chapter (Seo 2006). To begin with, there is a saying in Sub-Sahara, or even an old adage, that bears heavily with the topic of this chapter. That is, a man without a donkey is no man at all. The donkey, an important farm animal in the region, is an indicator of wealth and status in some societies in Sub-Sahara.

The Sahel is the lowland region below the Sahara Desert in Africa whose ecosystems are arid, semi-arid, and dry savannah. In Fig. 2.1, the region is marked by semi-arid and dry savanna zones below the Sahara Desert. The Central Africa is the lowland region below the Sahel whose ecosystems are sub-humid and humid forest zones, as marked by dark

S. N. Seo, *Climate Change and Economics*, https://doi.org/10.1007/978-3-030-66680-4_2

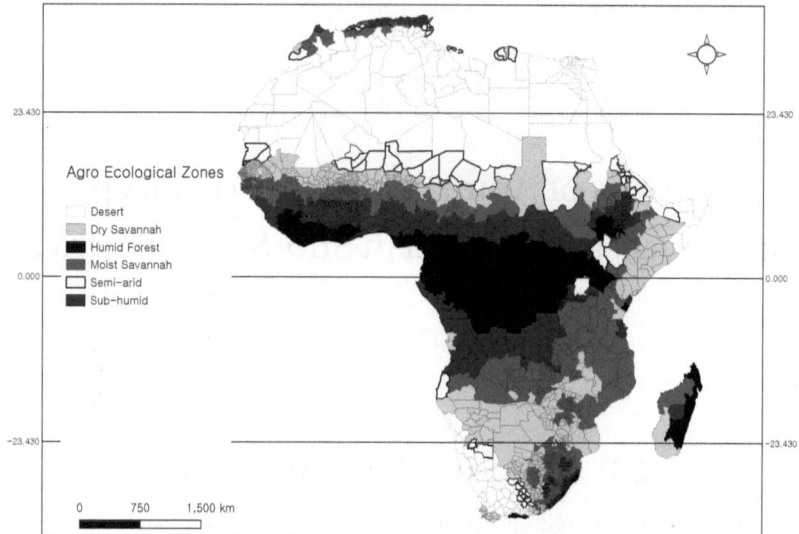

Fig. 2.1   Agro-Ecological Zones of Africa

colors in the figure. The Equator cuts through the Central African forest zones (FAO 2005; Seo et al. 2009a).

The moist savannah zones that lie between the Sahel and the Central African forest zones in the figure are roughly in the Sudanian savanna. In Central Africa, the River of Congo, a topic of another chapter in this book (Chapter 7), runs through Republic of Congo and the Democratic Republic of Congo from the Atlantic Ocean in the west to the Rwenzori National Park in the east.

The Sahara Desert is the largest desert in the world which splits Sub-Sahara from North Africa and the Mediterranean Sea. The highlands of Africa are located in the east of the continent where Mt. Kilimanjaro and the Serengeti are located, which can be verified in Fig. 2.2. Below Central Africa are mid-elevation ecological zones, except the eastern coastal zones where lowland ecological zones are dominant.

I believe that you, being a climate activist, have heard frequently about Sub-Sahara. This owes largely to two economic realities. One is that the continent is the poorest region in the world. The poverty rate is very high across the continent. There are about 520 million people who live under

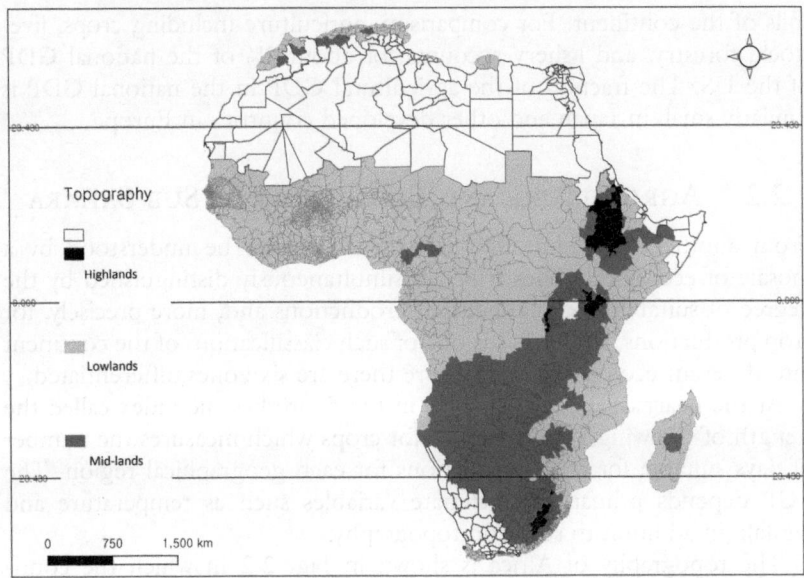

Fig. 2.2 Topography of Africa

extreme poverty in Africa, with the poverty rate amounting to nearly 40% of the total population (Refer to World Poverty Clock [WDL 2020]). The extremely high poverty rate is a cause of a high rate of malnutrition among children and women as well as a higher degree of vulnerability to various infectious diseases. The poverty problem in Africa remained unchanged despite other continents, for example, South Asia, have experienced a remarkable drop in the poverty rate during the late twentieth century, which I will come back to clarify in Chapter 4 (World Bank 2009).

The second reality is the continent's heavy reliance on agriculture as a source of income and livelihood. In many countries of Sub-Sahara, people employed by agriculture accounts for more than 80% of the total economic employment in these countries. In most Sub-Saharan countries, agricultural employment accounts for more than two-thirds (67%) of the total employment in the corresponding country (World Bank 2008).

This is an extremely high rate of reliance on agriculture which is again strongly influenced by outdoor conditions such as climate, weather, and soils of the continent. For comparison, agriculture including crops, live-stock, forestry, and fishery accounts for about 2% of the national GDP of the US. The fraction of the agricultural GDP in the national GDP is similarly small in Japan and other developed countries in Europe.

2.2   Agriculture and Ecosystems of Sub-Sahara

From another viewpoint, Sub-Saharan Africa can be understood by a mosaic of ecological zones which is simultaneously distinguished by the degree of suitability for agricultural productions and, more precisely, for crop productions. Figure 2.1 is one of such classifications of the continent into different ecological zones where there are six zones differentiated.

At the heart of the classification in the figure lies the index called the Length of Growing Periods (LGP) for crops which measures the number of days suitable for crop productions for each geographical region. The LGP depends primarily on climate variables such as temperature and rainfall, in addition to soils and topography.

The topography of Africa is shown in Fig. 2.2 in which the continent is distinguished by the three elevation designations: lowlands, mid-elevations, and highlands. It is noticeable that the Sahel, Sudanian savanna, and Central African forests are all in the lowland ecosystems. It is also salient that the highland areas are limited to eastern highlands in Mt. Kilimanjaro and in Stormberg-Drakensberg range in South Africa.

This classification is referred to as the Agro-Ecological Zone (AEZ) classification which had been developed as early as the late 1970s well before global climate challenges became a major international policy issue in the early 1990s through the Rio Earth Summit and the establishment of the United Nations Framework Convention on Climate Change (UNFCCC) (Dudal 1980; UNFCCC 1992).

According to the first of such classifications by Dudal who worked at the Food and Agriculture Organization (FAO) of the United Nations, the five AEZs were suggested based on the LGP concept (FAO 2005; Dudal 1980): desert, arid zone, semi-arid zone, sub-humid zone, and humid zone. To give you a rough idea of the LGP concept, be reminded that crops cannot be grown when temperature is below 5 degrees in Celsius; there is no sufficient precipitation; soils are not suitable for the crops grown.

As summarized in Table 2.1, the LGP is the number of days that is suitable for crop growth, as such, its range is from 0 to 365 days. The

Table 2.1 The Agro-Ecological Zones (AEZs) and the Length of Growing Period (LGP)

Continent	Agro-Ecological Zone (AEZ)	Length of Growing Periods (LGP)
Africa	Desert	0<=LGP<30 days
	Arid	30<=LG<90 days
	Semi-arid	90<=LGP<180 days
	Sub-humid	180<=LGP<270 days
	Humid	270<=LGP<365 days

desert has the LGP less than 30 days. The arid zone has the LGP greater than 30 days and smaller than 90 days. The semi-arid zone has the LGP greater than 90 days and fewer than 180 days. The sub-humid zone has the LGP greater than 180 days and fewer than 270 days. The humid zone has the LGP greater than 270 days and fewer than 365 days.

Which crops are most favored by African farmers? A survey of African farmers on their farming activities shows that the most favored crop is by far maize, also called corn in the US. Other than maize, other grains widely chosen include millet, sorghum, and wheat. Unlike South Asia, rice production is very limited. Other than these cereal crops, root crops, vegetables, and fruits and other tree products are cultivated in the continent. From the colonial periods, major cash crops that are exported include groundnut, coffee, cocoa, gum, and sugar (Seo et al. 2009b; World Bank 2009).

The basket of crop portfolios managed by African farms is summarized in Table 2.2. Maize is grown by 32% of African farms as a specialized portfolio. Another 17% of the farms grows maize as a diversified portfolio with fruits/vegetables. Another 14% of the farms grows maize as a diversified portfolio with groundnuts. A specialized millet farm accounts for 5% of the African farms. A specialized wheat farm accounts for 6% of the farms. A diversified portfolio of sorghum and millet accounts for 6% while a diversified portfolio of millet and groundnuts accounts for another 11%. A portfolio of fruits and vegetables, either specialized or diversified, accounts for another 10% of the African farms.

The table highlights the importance of maize in Africa agriculture and also explains why African development programs have invested so much into this crop (Byerlee and Eicher 1997). Of the total African farms surveyed, 62% of the farms raise maize. More strikingly, 32% of the

Table 2.2 The basket of African crop portfolios

Diversified or specialized?	Crop portfolios	Adoption percentages (%)
Specialized	Maize	31.67
Specialized	Millet	4.79
Specialized	Wheat	6.45
Diversified or Specialized	Fruits/Vegetables	9.94
Diversified	Maize and Fruits/Vegetables	16.62
Diversified	Maize and Groundnut	13.55
Diversified	Groundnut and Millet	10.86
Diversified	Millet and Sorghum	6.11

farms grow maize exclusively. The table also reveals the importance of the second cereal of the continent, that is, millet. About 22% of the farms are growing this grain crop. The millet is grown broadly with another major crop in a diversified crop portfolio.

2.3   FARM ANIMALS OF SUB-SAHARA

In addition to a large number of crops cultivated, Sub-Saharan farmers raise many different species of farm animals. To emphasize, the basket of portfolios summarized in Table 2.2 is revealing information only about crops, setting aside farm animals. The same survey of African farmers shows that the most favored livestock species in the continent are beef cattle, dairy cattle, sheep, goats, and chickens and the less frequently adopted species include pigs, donkeys, horses, beehives, guinea fowls, and so on (Seo and Mendelsohn 2008).

The farm animals are of interest to us because Sub-Saharan Africa is vastly occupied by desert, arid zones, and semi-arid zones, all of which are not "ideal" places for crop cultivation. On the other hand, these agro-ecological zones are most often suitable or even sometimes optimal for raising farm animals, as long as there is sufficient rainfall that can support grasses and pasture. The Sahelian arid zones, depicted in Fig. 2.1, therefore, may turn out to be an advantageous place for animal husbandry.

To examine this hypothesis, in Fig. 2.3, I summarize the average number of each farm animal species per farm in the farms that chosen the animal species (Seo 2014). The five most frequently chosen animals in Sub-Sahara are beef cattle, dairy cattle, sheep, goats, and chickens. The figure indicates that our hypothesis is true: The number of farm animal species per each farm is far greater in the arid climate zones, especially so for goats and sheep.

Across the sheep farms, the average number owned is as large as 449 in the desert zone and 39 in the arid zone. By contrast, the average number of sheep owned is only 3 in the humid zone. The semi-arid zone owns about 11 sheep per each farm, so does the sub-humid zone.

Across the goat farms, the average number of goats owned is as large as 59 goats per farm in the desert zone and 39 goats per farm in the arid zone. By contrast, the humid zone has only 3.7 goats per farm. The semi-arid zone has 7.6 goats and the sub-humid zone has 4.8 goats per farm.

A similar distribution is observed for dairy cattle, although with smaller differences in the numbers of dairy cattle across the zones: 10 dairy cattle for the desert zone, 7 for the arid zone, 3.3 for the semi-arid zone, 2.7 for the sub-humid zone, and 0.8 for the humid zone.

A similar distribution is not observed for either beef cattle or chickens. Especially for chickens, a reverse distribution is found. The chicken farms in the humid zone own 149 chickens per farm while the chicken farms in the desert zone own only 4 chickens per farm. There are 11 chickens per farm in the arid zone, 22 chickens in the semi-arid zone, and 38 chickens in the sub-humid zone. Notice the reverse trend: the wetter the region, the larger the average number of chickens per farm.

For beef cattle, the distribution is somewhat similar to that of dairy cattle. But, unlike dairy cattle, the number of beef cattle raised in the sub-humid zone is as large as 9 and that in the semi-arid zone is as large as 11.

2.4   LIVESTOCK MANAGEMENT AND CLIMATE CHANGE IN SUB-SAHARA

Figure 2.3 indicates that our hypothesis may be true: Even though arid and semi-arid zones may not be optimal places to grow crops, these zones may be advantageous for raising farm animals relying on grasses and the mobility of animals to seek water and find shades. What the figure reveals

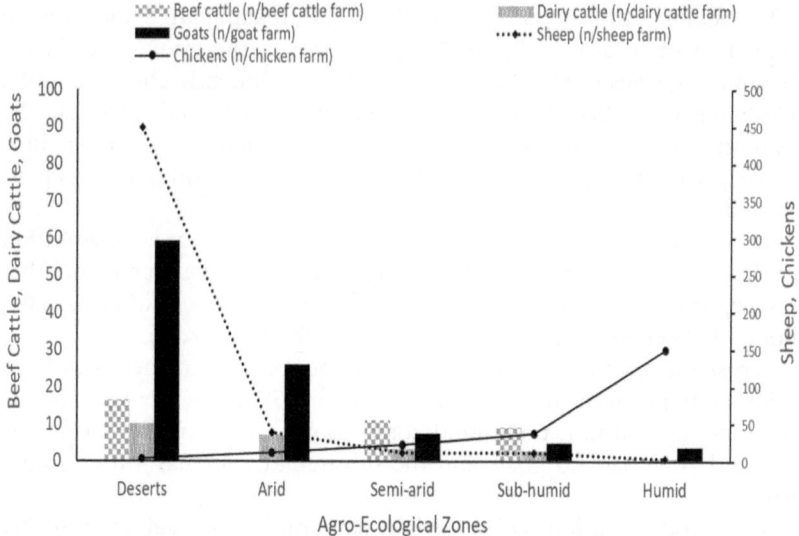

Fig. 2.3 Farm animals and ecosystems in Sub-Sahara: average numbers of animals owned

is that the more arid the zone is, the larger the numbers of goats, sheep, and cattle raised by Sub-Saharan farmers are.

The raising of farm animals is referred to as livestock management, or animal husbandry for a small-scale family-based operation, which is regarded as a component of agriculture. We can analyze the numbers presented in Fig. 2.3 more rigorously with statistical techniques, which would enable us additionally to predict the impacts of climate change on the numbers of animals in the decades to come. Will a warmer Planet lead to a devastation of animal farms? This is just one of the questions that can be answered with the statistical analyses.

Livestock management is an essential component of agriculture in any continent and any country of the Planet. To offer you some evidence on this, of all the farms in the US, 53% of them own at least some farm animals. Of the total agricultural income generated in the US, the income from livestock management accounts for 49% (USDA 2007). In South America as well as Africa, at least two-thirds of the farms manage at least some farm animals (Seo and Mendelsohn 2008; Seo 2010).

Of the range of the statistical modeling efforts on African agriculture, let me introduce a probabilistic model of choice of farm animals which enables researchers to estimate the probability of each farm animal portfolio being chosen under various climate conditions. The probabilistic model will also enable us to identify the effects of climate change on the choices of animal portfolios.

In Fig. 2.4, I draw four hypothetical choice probability distributions across the range of temperature normal, which captures the core of the economic theory of adaptation. The temperature normal is a 30-year average temperature widely used in the climate science, so is different from a single year average temperature or a single day temperature. The four distributions capture the following:

> *Species I: Preferred in colder temperature;*
> *Species II: Preferred in moderate temperature;*
> *Species III: Preferred in hotter temperature;*
> *Species IV: Preferred in hottest temperature.*

Notice that the four distributions have four different functional forms, that is, shapes: roughly speaking, a linear shape, a hill-shape, logarithmic,

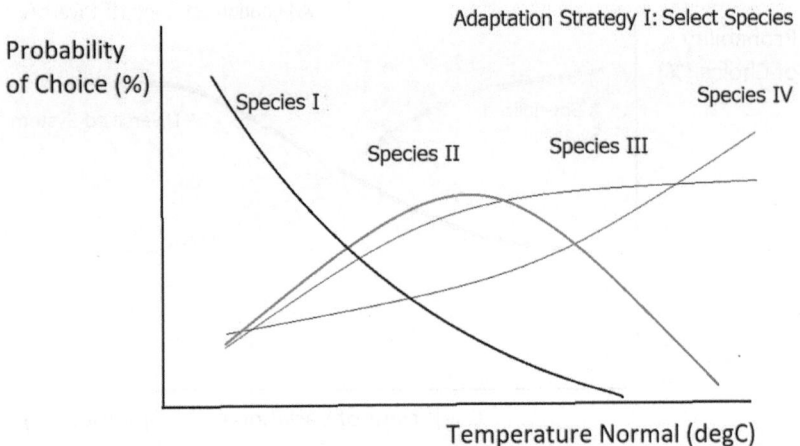

Fig. 2.4 Hypothetical choice probabilities of farm animals across temperature normal

a power-law shape. The most salient feature of Fig. 2.4 is that some animal species are increasingly favored in higher temperature zones while some animals are favored in moderate temperature or colder temperature. This feature is underneath the theory of adaptation behaviors.

The four hypothesized distributions were in fact found in the study of Sub-Saharan farms (Seo and Mendelsohn 2008). Resembling species IV in the figure, the probability of adopting goats increases as the temperature normal gets hotter. The probability of adopting sheep also increases as the temperature normal gets higher, but in the distribution of species III. The probability of adopting beef cattle as well as that of dairy cattle falls, the hotter the temperature normal becomes, as in the distribution of species I. There is little beef cattle raising when the temperature normal exceeds 26 °C. For chickens, the probability of adoption increases initially, reaches a peak, and falls afterward. The peak is at around 21 °C in temperature normal. This distribution is similar to species II in the figure.

A similar statistical analysis can be applied to study farm choices across the range of precipitation normal in Sub-Sahara. This is of high pertinence since the aforementioned arid zones, semi-arid zones, and other AEZs are defined in terms primarily of precipitation. Figure 2.5 shows two hypothesized choice probability distributions across precipitation risk.

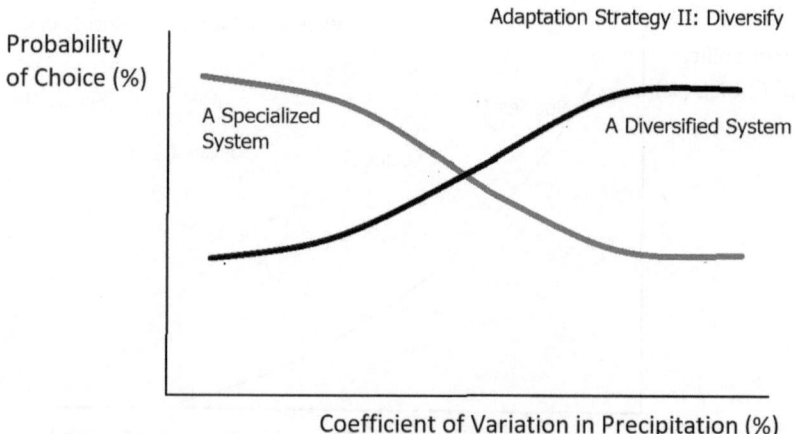

Fig. 2.5 Hypothesized choice probability distributions across precipitation risk

Precipitation risk is captured by the variability of precipitation for the 30-year period, more concretely, the coefficient of variation in precipitation (CVP) which is the degree of variability instead of the absolute size of variability.

Two distributions are labeled as a specialized system and a diversified system of agriculture. The diversified system's distribution shows an increasing probability as precipitation risk increases up to a plateau probability. The specialized system's distribution is the opposite: It shows a declining probability of choice as precipitation risk increases down to a floor probability. Summing up,

The specialized system: Preferred in lower precipitation risk zones;
The diversified system: Preferred in higher precipitation risk zones.

The two hypothetical distributions are in fact found in Sub-Saharan agriculture. The specialized system's distribution resembles that of the specialized crop system of agriculture. The diversified system's distribution approximates that of the diversified crop-livestock system of agriculture of the continent (Seo 2012).

The precipitation risk in Sub-Sahara is, among other factors, determined by the multi-decadal shifts in the ocean currents in the North Atlantic Ocean called the Atlantic Multi-decadal Oscillation (AMO) (Janowiak 1988). Owing to the AMO, the CVP is especially high in the Sahelian arid and semi-arid zones (Seo 2012).

What are the reasons for the farmers' preference of farm animals over crops in the arid and semi-arid zones? One of the well-reported mechanisms is an increased probability of an outbreak of an infectious livestock disease in high rainfall and humid conditions (Ford and Katondo 1977). A vector-borne disease, the trypanosomiasis or more commonly called the sleeping sickness, carried by tsetse flies is the most prominent of such diseases and has received much attention.

Another well-reported mechanism is the interrelationship between livestock management and crop farming. In the arid zones, crop farming is grueling, if not impossible, for many water-intensive crops such as maize (corn), rice, and wheat. During the planting and growth periods of these crops, crop fields should be filled with water at an ankle-high level (Welch et al. 2010). Even in these zones, however, animals can be raised as long as there are pastures or grasses which can be sustained with only small amount of rainfall or irregular rainfalls. Unlike the crops which require an irrigation facility in dry zones, farm animals can seek water from nearby rivers and lakes.

2.5   KEY TAKEAWAYS

You might have wondered, before reading this chapter, why the author of this book begins the textbook on climate change, economy, and livelihood with a story of farm animals in Sub-Sahara. I hope you feel, having read this chapter through, that the story is a fascinating showcase of the inter-relationships among climate change, economy, and individuals' livelihood decisions. In fact, the field of the economics of adaptation behaviors to climate change was originated from this research 20 years ago by the present author and numerous collaborators, especially, the World Bank and Yale University (Seo 2006).

Of the many insights that can be gleaned from this research, the key take-home message is that the climate system has always been an important factor in the economic decisions in some economic sectors, agricultural and natural resource uses in particular, and agricultural/resource managers have long sensibly coped with the changes in the climate system by employing a host of strategies and practices.

The two aspects of the key takeaway, although described as if common sense in this chapter, are often looked over and set aside in the climate change reports and policy roundtables (IPCC 1990, 2021). For the past 30 years of publications from the Intergovernmental Panel on Climate Change (IPCC), there has been little attention laid upon an economic analysis while a heavy emphasis has been directed to climate science (Mendelsohn 2016). Another way to state this is that the empirical results described, as well as theoretical models implied, in this chapter are subtle and difficult to obtain elsewhere, that is, without a solid economic analysis and subsequently heavy efforts.

2.6   THE FULL PICTURE

The story of farm animals told up to this point, however fascinating as it may be, does not end here. The complete picture of the story on farm animals and climate change must bring up another key linkage: the effects of farm animals on greenhouse gas emissions. As alluded before, farm animals are a source of methane emissions, a potent greenhouse gas, which must be portrayed in the full picture.

The methane emissions from all sources may account for about 25% of the greenhouse gas emissions globally, of which the emissions from livestock management is the largest anthropogenic source (GCP 2016). The story of livestock methane emissions simultaneously brings our attention to greenhouse gas reduction strategies in the livestock and agricultural sector. Put slightly differently, the livestock sector offers multiple opportunities to reduce the greenhouse emissions for the benefit of the Planet, for example, by developing alternative feeds to livestock.

We will come back to this topic in Chapter 6 devoted to the story of majestic grasslands on the Planet. The chapter will be one of the chapters in the book devoted to explaining a variety of mitigation strategies.

2.7 Chapter Highlights

- When it comes to the impacts of climate change, Sub-Saharan Africa has received most attention owing to the continent's already adverse climatic conditions.
- In particular, agriculture, the primary economic sector of the continent, has long been believed and argued to be highly vulnerable to future climatic changes.
- This chapter shows that a closer examination of Sub-Saharan agriculture reveals a great diversity of agricultural activities whose distributions are associated with a range of agro-ecological zones.
- Besides the most-studied crops of Sub-Sahara, farm animals are an essential component of Sub-Saharan agriculture especially in arid, semi-arid, and savanna zones. The most frequently owned animals are beef cattle, dairy cattle, goats, sheep, and chickens.
- An economic analysis shows that Africa farmers prefer to adopt some farm animals in a hotter climate zone, especially goats and sheep. Four different distributions in the choice probability of a farm animal are found.
- In response to higher climate risk through higher precipitation variability, African farmers switch to an integrated crop-livestock system, moving away from a specialized crop system.
- A complete picture of adaptation to climate change cannot be painted until Sub-Saharan farmers would figure out the strategies to reduce methane emissions from farm animals, a key topic for later chapters of this book.

REFERENCES

Byerlee, D., and C.K. Eicher. 1997. *Africa's Emerging Maize Revolution.* Boulder, CO: Lynne Rienner Publishers Inc.

Dudal, R. 1980. *Soil-Related Constraints to Agricultural Development in the Tropics.* Los Banos, The Philippines: International Rice Research Institute.

Food and Agriculture Organization (FAO). 2005. *Global Agro-ecological Assessment for Agriculture in the Twenty-first Century (CD-ROM), FAO Land and Water Digital Media Series.* Rome: FAO.

Ford, J., and K.M. Katondo. 1977. Maps of Tsetse Fly (Glossina) Distribution in Africa, 1973, According to Subgeneric Groups on a Scale of 1: 5000000. *Bulletin of Animal Health and Production in Africa* 15: 187–93.

Global Carbon Project (GCP). 2016. Global Methane Budget 2016. GCP. Accessed from https://www.globalcarbonproject.org/methanebudget/index.htm.

Intergovernmental Panel on Climate Change (IPCC). 1990. *Climate Change: The IPCC Scientific Assessment.* Cambridge: Cambridge University Press.

Intergovernmental Panel on Climate Change (IPCC). 2021. *Climate Change 2021: The Physical Science Basis, The Sixth Assessment Report of the IPCC.* Cambridge: Cambridge University Press.

Janowiak, John E. 1988. An Investigation of Interannual Rainfall Variability in Africa. *Journal of Climate* 1 (3): 240–55.

Mendelsohn, Robert. 2016. Should the IPCC Assessment Reports Be an Integrated Assessment? *Climate Change Economics* 07 (01): 1640002.

Seo, S. Niggol. 2006. Modeling Farmer Responses to Climate Change: Climate Change Impacts and Adaptations in Livestock Management in Africa. PhD dissertation, Yale University, New Haven, CT.

Seo, S. Niggol. 2010. A Microeconometric Analysis of Adapting Portfolios to Climate Change: Adoption of Agricultural Systems in Latin America. *Applied Economic Perspectives and Policy* 32 (3): 489–514.

Seo, S. Niggol. 2012. Decision Making Under Climate Risks: An Analysis of Sub-Saharan Farmers' Adaptation Behaviors. *Weather, Climate, and Society* 4 (4): 285–99.

Seo, S. Niggol. 2014. Evaluation of the Agro-Ecological Zone Methods for the Study of Climate Change with Micro Farming Decisions in Sub-Saharan Africa. *European Journal of Agronomy* 52 (January): 157–65.

Seo, S. Niggol, and Robert Mendelsohn. 2008. Measuring Impacts and Adaptations to Climate Change: A Structural Ricardian Model of African Livestock Management. *Agricultural Economics* 38 (2): 151–65.

Seo, S. Niggol, Robert Mendelsohn, Ariel Dinar, Rashid Hassan, and Pradeep Kurukulasuriya. 2009a. A Ricardian Analysis of the Distribution of Climate Change Impacts on Agriculture Across Agro-Ecological Zones in Africa. *Environmental & Resource Economics* 43 (3): 313–32.

Seo, S. Niggol, Robert Mendelsohn, Pradeep Kurukulasuriya, Ariel Dinar, and Rashid Hassan. 2009b. Differential Adaptation Strategies to Climate Change in African Cropland by Agro-Ecological Zones. Policy Research Working Paper #460, World Bank, Washington, DC.

United Nations Framework Convention on Climate Change (UNFCCC). 1992. *United Nations Framework Convention on Climate Change*. New York, NY: UNFCCC.

United States Department of Agriculture (USDA). 2007. *Census of Agriculture 2007*. Washington, DC: USDA. Accessed from http://www.agcensus.usda.gov/Publications/2007/index.php.

Welch, Jarrod R., Jeffrey R. Vincent, Maximilian Auffhammer, Piedad F. Moya, Achim Dobermann, and David Dawe. 2010. Rice Yields in Tropical/Subtropical Asia Exhibit Large but Opposing Sensitivities to Minimum and Maximum Temperatures. *Proceedings of the National Academy of Sciences* 107 (33): 14562–67.

World Bank. 2008. *World Development Report 2008: Agriculture for Development*. Washington, DC: World Bank.

World Bank. 2009. *Awakening Africa's Sleeping Giant: Prospects for Commercial Agriculture in the Guinea Savannah Zone and Beyond*. Washington, DC: World Bank and FAO.

World Data Lab (WDL). 2020. World Poverty Clock. Accessed from https://worldpoverty.io/ on November 3, 2020.

CHAPTER 3

Giving Forests: A Tale of Amazon Rainforests and Congo River Forests

3.1 FORESTS AND GLOBAL WARMING

Forestry, unlike farm animals in the previous chapter, has been placed at the central stage of climate change science as well as climate policy negotiations from the very beginning. From the angle of a policy focal point with regard to global warming, deforestation, that is, cutting trees for timber sales or crop farming, releases carbon dioxide into the atmosphere. A rapid deforestation in the Amazon rainforests and the Indonesian forests has thus received much attention as a source of carbon dioxide emissions.

The intense interests on Amazon deforestation, however, preceded the global warming implications (Peters et al. 1989). Many environmentalists identified the rapid deforestation as the evidence of the side-effects of rapid economic development. The deforestation meant to them the loss of valuable natural resources of the Planet, which is accompanied by a host of negative consequences including losses of habitats for animals and birds, loss of soils, loss of livelihoods of local people.

Another group of thinkers points out the socio-economical complexities of the rapid deforestation in the Amazon. A deforestation is unavoidable for a remote locality's economic development and well-being because roads must be constructed to connect to nearby urban areas. A slash-and-burning of forests may be called for in certain communities for a cropping area to be established or for obtaining fuel woods. In another situation,

S. N. Seo, *Climate Change and Economics*, https://doi.org/10.1007/978-3-030-66680-4_3

43

forest areas may be the sites of valuable petroleum deposits, in which case the local community must clear the trees for oil and gas revenues, e.g., the Yasuni reserve in Ecuador.

Having said that, the interests in the Amazon rainforests may be even beyond the environmental and sociological concerns. The Amazon rainforest has long been recognized as one of the Planet's most valuable treasures. The Europeans had long looked for the El Dorado, the city of gold, there. It is the world's largest rainforest. The number of local communities that live inside the Amazon rainforest with their unique cultures may be unaccountable. Rare species of animals and plants that are only found in the Amazon may be irreplaceable.

With these tall and rich tales in the milieu, the mysterious and mighty Amazon rainforest has added in the recent decades another mystique to it: global warming. The second and third largest rainforests are the Congo River rainforests in Sub-Sahara and the Indonesian rainforests. You can verify the three great forests in Fig. 3.1 (Matthews 1983).

It goes without saying that the Amazon rainforest is not the only great forests in the Planet to which much attention has been directed. Broadly, there are boreal forests in the high latitude regions, say, cold deciduous

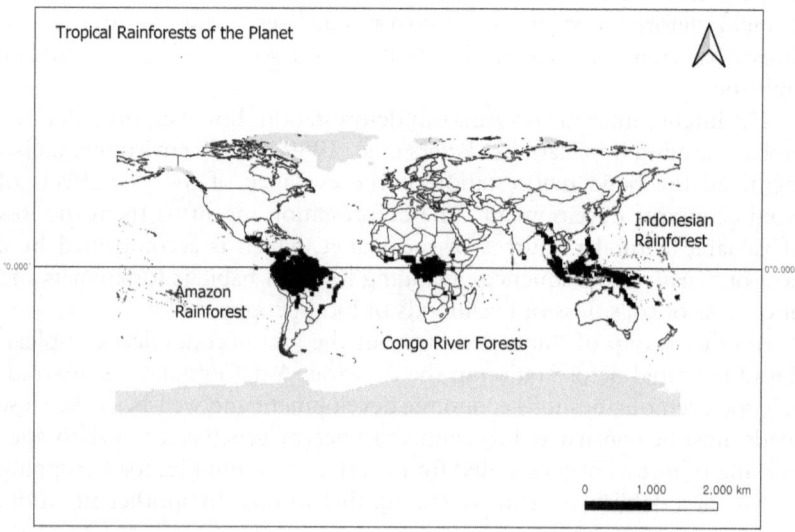

Fig. 3.1   Rainforests

forests such as Douglas fir and Norwegian spruce. Also, the austral forests in the Southern Hemisphere are prominent, which hosts many sclerophyllous (meaning thick-leaved) trees such as Eucalyptus (snow gum) and Fig (Banyan). Even in the dry zones, trees survive, e.g., xerophytes such as Joshua tree.

According to a recent research, the total number of individual trees on the Planet is estimated to be about 3.04 trillion, of which approximately 1.30 trillion exist in tropical and subtropical forests, 0.74 trillion in boreal regions, and 0.66 trillion in temperate regions (Crowther et al. 2015). Annually, the authors estimate that 15 billion trees are cut down for numerous economic purposes.

In this chapter, I will tell the global warming story of forests with a focus on the Amazon rainforest in Latin America and the Congo River rainforests in Sub-Sahara. Again, the story will highlight the intricate inter-relationships among climate change, economic activities, and livelihoods of people.

3.2   Amazon Rainforest

The Amazon rainforest is the world's largest rainforest, accounting for about half of the Planet's rainforest. The Amazon basin to which the rainforest belongs is as large as 6.3 million square kilometers, which is nearly 18 times larger than the land territory of Japan or as large as two-thirds of the US. The Amazon basin refers to the geographical area in South America drained by the Amazon River and its tributaries.

The Amazon rainforest belongs to nine countries in South America: Brazil (60%), Peru (13%), Colombia (10%), Venezuela, Ecuador, Bolivia, Guyana, Suriname, and French Guiana. The South American continent that contains the Amazon is the most densely-forested continent on Earth with 44% of its land area covered by the densely forested zones defined as the area with greater than 50% coverage. For comparison, the densely-forested zones account for 36% of total land area in Sub-Saharan Africa (WRI 2005).

According to a research by the Field Museum of Chicago, there are as many as 16,000 tree species and 390 billion individual trees in the Amazon rainforest (ter Steege et al. 2013). Another group of researchers mentioned previously indicate that the number is still an underestimate (Crowther et al. 2015). Of the 16,000 tree species, it is estimated that

227 species are a "hyperdominant" species, accounting for about 50% of the total number of individual trees.

The 20 most abundant tree species, that is, hyperdominant species, in the Amazon rainforest are as follows, with scientific names. Their common names or most-widely known species are added by the present author inside the parentheses (ter Steege et al. 2013):

- Euterpe precatoria (a tall palm native to Central and South America);
- Protium altissimum (Burseraceae, the torchwood family including candlewood tree, copperwood tree);
- Eschweilera coriacea (a Lecythidaceae family, including Brazil nut);
- Pseudolmedia laevis (commonly known as lechechiva; a mulberry family);
- Iriartea deltoidei (an evergreen single-stemmed palm tree);
- Euterpe oleracea (the acai palm);
- Oenocarpus bataua (the patawa, a palm tree);
- Trattinnickia burserifolia (a native South American tree known as Amesclao in Brazil, Pulgande in Ecuador);
- Socratea exorrhiza (the walking palm with stilt roots);
- Astrocaryum murumuru (a palm with murumuru butter);
- Brosimum lactescens (a Moraceae family, the mulberry or fig family);
- Protium heptaphyllum (an evergreen and ornamental tree);
- Eperua falcata (a large evergreen tree);
- Hevea brasiliensis (commonly known as the Para rubber tree);
- Eperua leucantha (an evergreen tree belonging to the Legume family);
- Helicostylis tomentosa (a Moraceae family, the fig, banyan, mulberry family);
- Attalea butyracea (a palm tree native to Mexico);
- Rinorea guianensis (Violaceae family, including violets and pansies);
- Licania heteromorpha (Chrysobalanaceae family, including coco plum);
- Metrodorea flavida (a Rutaceae family, including citrus).

According to another source, other than the dominant tree species, notable plants and trees of the Amazon are as follows in their family names with their rough common names or representative species shown inside the parentheses (Giacometti 1990):

- Annonaceae (commonly known as the custard apple family);
- Apocynaceae (dogbane family);
- Arecaceae (commonly known as palms);
- Bromeliaceae (a family of monocot flowering plants including Spanish moss, pineapple, succulents);
- Chrysobalanaceae (a family of flowering plants, including coco plum);
- Clusiaceae (a family of trees and shrubs, with milky sap and capsules for seeds);
- Euphorbiaceae (Euphorbia, the Spurge family including cassava, castor oil plant, Para rubber tree);
- Fabaceae (commonly known as the Legume, pea, bean family);
- Heliconiaceae (commonly known as lobster-claws, toucan beak, wild plantains, or false bird-of-paradise);
- Lecythidaceae (a family of woody plants including Brazil nut, paradise nut);
- Malpighiaceae (a family of flowering plants, including acerola cherry);
- Malvaceae (the mallows, including cacao, cotton, durian, okra);
- Meliaceae (the mahogany family);
- Myrtaceae (the myrtle family, including myrtle, guava, eucalyptus);
- Olacaceae (a family of woody plants native to tropical regions);
- Orchidaceae (the orchid family);
- Rubiaceae (commonly known as the coffee family).

I hope the readers were able to identify some of the trees, plants, and flowers in the above lists. Conspicuously, there are many different species of the palm tree in the Amazon rainforest. Also, it is apparent that California redwood and Eucalyptus, favorites of many people, are not found there often.

3.3   Carbon Cycle

The relationship between forests and climate change and its central position in the climate change is best captured by the theory of carbon cycle (Schlesinger 1997). The carbon cycle is a description of the carbon exchanges among the many reservoirs of carbon on Earth. The reservoirs are atmosphere, oceans, soils, fossil fuels, vegetations, microbes, rivers, permafrost, and humans. The ocean reservoir is further divided into surface ocean, deep ocean, ocean floors, and marine biota.

The carbon cycle is depicted in Fig. 3.2. Anthropogenic activities emit annually 7.8 gigatons of carbon (GtC) through fossil fuel uses and cement production and additionally 1.1 GtC through net land use changes. From the emission of 9 GtC per year, forests absorb as much as 4.3 GtC from the atmosphere annually, that is, the difference between photosynthesis and respiration shown in the figure (Refer to the land box). Additionally, oceans absorb 1.6 GtC annually, that is, the difference between the release and the sink shown in the figure (Refer to the ocean box) (Ciais et al. 2013).

Of the aforementioned reservoirs, a reservoir that stores carbon is called a carbon sink. A forest/plant reservoir is therefore a carbon sink. Another is soils. The third is the oceans. The fourth is fossil fuels. As depicted in Fig. 3.2, the forest/plant reservoir, labeled the vegetation in the figure, stores about 550 GtC while the soil reservoir stores another 2,000 GtC. A fossil fuel reservoir consisting of oil, gas, and coal stores

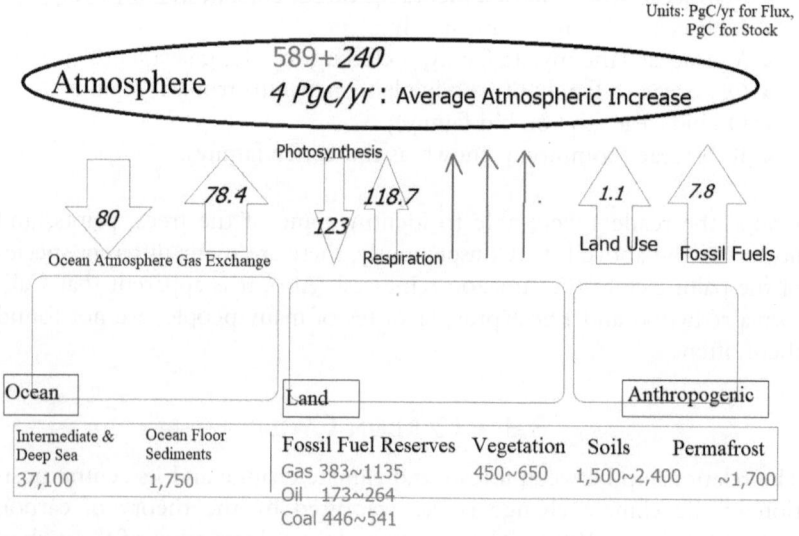

Fig. 3.2 Carbon cycle and forests

10,000 GtC. The ocean reservoir stores as much as 40,000 GtC in surface ocean, deep ocean, ocean floor, and marine biota.

The forest/plant reservoir provides the largest sink of carbon annually. The mechanism is known as the photosynthesis. It is a process used by plants and trees to convert light energy to chemical energy. The chemical energy is stored in the form of carbohydrate molecules, which is later released to power activities of organisms. For the photosynthesis, plants and trees synthesize carbon dioxide and water with light energy. As a byproduct, the process produces oxygen. Besides trees and plants, most algae and cyanobacteria perform photosynthesis.

Succinctly, the photosynthesis by plants on Earth is the following chemical process (Ciais et al. 2013):

$$6CO_2 + 6H_2O \xrightarrow{Sunlight\ (photons)} C_6H_{12}O_6 + 6O_2. \tag{3.1}$$

The photosynthesis process is a profoundly important process to humanity and life on Earth, to say the least. This is witnessed by the right side of the process in Eq. 3.1. It provides and maintains the oxygen content of the Planet. It provides most of the energy necessary on Earth to sustain life of humans and other sentient beings by producing glucose, carbohydrates.

As far as global warming is concerned, the photosynthesis process is the mechanism of carbon sink by plants and trees. This is witnessed by the left side of the process in Eq. 3.1. Carbon dioxide in the atmosphere is converted, therefore absorbed, through the photosynthesis to plant biomass. Annually, the forest/plant reservoir absorbs as much as 123 GtC annually as you can verify in Fig. 3.2.

The CO_2 captured by the reservoir is released into the atmosphere again through respirations of plants and forests. Vegetation fires also release CO_2 and methane (CH_4) into the atmosphere. As depicted in the figure, the amount of release from plant respirations and fires is 119 GtC per year. The respiration is the following chemical process:

$$C_6H_{12}O_6 + 6O_2 \rightarrow 6CO_2 + 6H_2O + \text{heat}. \tag{3.2}$$

3.4   CARBON EMISSIONS
FROM DEFORESTATION AND LAND USE CHANGES

Another process that releases carbon from the terrestrial ecosystems is land use changes, of which deforestation is the primary anthropogenic activity. This is marked as the arrow above the "Anthropogenic" box in Fig. 3.2. Land use changes are defined as changes in land uses and covers, which include, *inter alia*, deforestation, afforestation, reforestation, desertification, conversions of agricultural lands, and changes in urban land uses. Hence, an increase in carbon emission from the land use changes can be largely attributed to changes in the forest/plant ecosystems.

The carbon emissions from land uses has been declining. The estimate of carbon emissions from the land use changes was 1.4 GtC/year during the 1980s, 1.5 GtC/year during the 1990s, 1.1 GtC/year during the 2000s, and 0.9 GtC/year during the 2100s. These are global estimates, which are obtained from different scientific methods: a book-keeping method, a terrestrial ecosystem modeling, and a satellite data analysis (Houghton et al. 2012; Harris et al. 2012).

Land use emissions of carbon dioxide are quite varied across the world regions. The net emissions of carbon dioxide are concentrated in the continents and countries where rich forest resources are located. These countries supply timber and wood products to the world, as such, deforestation rates thereof are high. Figure 3.3 shows land use carbon dioxide emissions across the world regions, which reveals that the emissions are concentrated in Latin America and South Asia (Houghton 2008).

Per decade during the 2000s, the amount of carbon emissions from changes in land use is as high as 6 GtC in Latin America, which corresponds to about 23 gigatons of carbon dioxide ($GtCO_2$). It is also high with 6.2 GtC per decade in South Asia where major timber exporting countries such as Indonesia are located. The amount of carbon emissions is 2.6 GtC for Africa or 9.5 $GtCO_2$. For your reference, the formula to covert the amount of carbon to the amount of carbon dioxide is to multiply the former by 44/12.

Notably, the amount of land use emissions is negative or near zero for the developed regions: the US, EU, and Japan. It is also notable that the amount of emissions from China and India are negative or near zero. In these regions, an increase in vegetations and forests owing to afforestation and reforestation outpaces a decrease in vegetations and forests owing to deforestation and other changes.

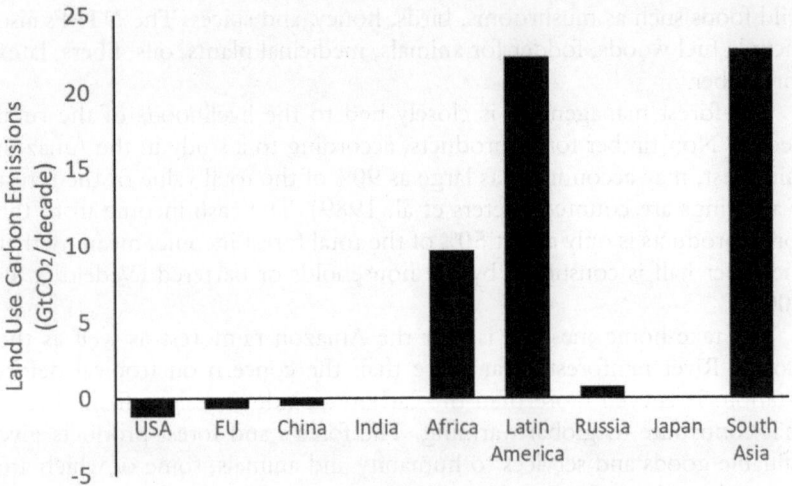

Fig. 3.3 Land use emissions of World Regions

3.5 Economics: Forest Income and Livelihoods

Up to this point, my stories in this chapter were limited largely to the scientific aspects of climate change and forests, say, forest coverage, carbon cycle, photosynthesis, and carbon emissions. From this point on, I will highlight the economic and livelihood aspects of forests under changes in the climate system.

A meta-analysis of the rural poor shows that forest income is a significant source of income for them. It accounts for as much as 22% of the rural income globally, taking into consideration cash income, own consumption, and barter. Regionally, the forest income represents as much as 35% of the rural income in Latin America, the focal point of this chapter. In Asia, the forest income represents 18% while, in Southern Africa, 25% of the total rural income is accounted for by the forest income (Vedeld et al. 2004). In the Congo River basin in Central Africa, the percentage is likely even higher than that of Southern Africa.

The forest income is earned from two types of products: timber and non-timber forest products (NTFP) (Peters et al. 1989). The timber production includes both sawlogs and pulpwood. The NTFPs are numerous services and products that forests offer other than timber. The NTPFs include edible fruits such as cacao, brazil nut, and persimmon;

wild foods such as mushrooms, birds, honey, and spices. The NTPFs also include fuel woods, fodder for animals, medicinal plants, oils, fibers, latex for rubber.

The forest management is closely tied to the livelihoods of the rural people. Non-timber forest products, according to a study in the Amazon rainforest, may account for as large as 90% of the total value of the forest if all things are counted (Peters et al. 1989). The cash income from the forest products is only about 50% of the total forest income, meaning that the other half is consumed by the households or bartered (Vedeld et al. 2004).

The take-home message is that the Amazon rainforest as well as the Congo River rainforest is far more than the concern on tropical deforestation. It is even more than the carbon dioxide emissions from forests that contribute to global warming. The forests and forest products give valuable goods and services to humanity and animals, some of which are essential products that can be sold and consumed by the forest managers for sustaining their lives.

3.6   ECONOMICS: FOREST GROWTH UNDER ELEVATED CO_2

The increase in carbon dioxide in the atmosphere enhances the process of photosynthesis for trees and other vegetations, as explained above. Put differently, the enhanced photosynthesis coverts carbon dioxide to plant biomass in trees. As such, trees, shrubs, grasses are expected to grow taller and thicker.

Such predictions are made by climate and ecosystem scientists through either a laboratory-based experiment or a field-based experiment. The latter is often referred to as the FACE experiment, short for the Free Air Carbon Enrichment experiment, which is more realistic than the former. You may remember that the FACE was mentioned in Chapter 1. In the FACE experiment, a plot of forest land is chosen for experiments and the level of carbon dioxide in the neighborhood of the plot is enhanced via the supply of the molecule through the pre-installed pipes around the plot.

The FACE experiments have been conducted since the early 1990s (Ainsworth and Long 2005). A meta-analysis of the results from numerous studies for a 15-year period is summarized in Fig. 3.4, concerning the growth of trees and other vegetations. The carbon dioxide

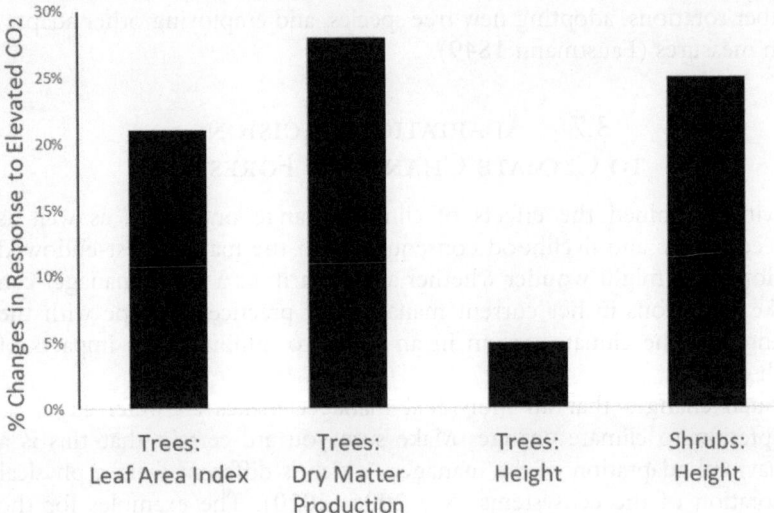

Fig. 3.4 Impacts of carbon dioxide doubling on trees by the FACE

enhancement effects on trees are substantial: 21% increase in Leaf Area Index (LAI), 28% increase in Dry Matter Production (DMP), and about 5% increase in height. The effects on shrubs are also substantial: about 25% increase in height. These changes occurred on average in all the FACE studies during the 15-year period in response to a doubling of carbon dioxide concentration in the neighborhood of the experimental plots.

At this point, you might be wondering: if individual trees would grow taller and faster, what would happen to the landscape of trees you are familiar with? You are right. The distribution of trees and shrubs you are accustomed to today will be shifted to another one because of the above-described effects on individual trees. This redistribution would occur not only at the national level but also at the Planet level (Joyce et al. 2000). The colorful red maples that light up New England towns in the fall may move out across the border to Canada.

The redistribution of trees at such large geographical scales will have an important implication on forest managers and the forest industry (Sohngen and Mendelsohn 1998). An optimal management of forest resources will need to take into account such redistributions by altering timber rotations, adopting new tree species, and employing other adaptation measures (Faustmann 1849).

3.7   ADAPTATION DECISIONS
TO CLIMATE CHANGE IN FORESTRY

Having examined the effects of climate change on forests as well as the economic and livelihood consequences in the major forest-endowed regions, you might wonder whether a rural farm or a forest manager can make alterations in her current management practices to cope with the changes in the climate system in an effort to minimize the impacts of such changes.

Such changes that an individual manager makes are referred to as adaptation to climate change. Make sure you are certain that this is a behavioral adaptation of the managers which is different from a physical adaptation of the ecosystems (Seo 2006, 2020). The examples for the latter are: trees can adapt to a drier climate by learning to store and use water more efficiently, as in the xerophytes; an animal may become more heat tolerant as it gets used to a hotter climate. Of the behavioral adaptations, efficient adaptations are behavioral changes that choose the most profitable alternative of all available options (Mendelsohn 2000).

A study in Sub-Saharan Africa which included the Congo River rainforest identified that the variety of farming activities performed there can be grouped into four systems: a crops-only system, a crops-livestock system, a crops-forests system, a crops-livestock-forests system (Seo 2010). Of the four, you can see that the two systems mix forest-related economic activities with other farming activities.

In Fig. 3.5, I summarize adoption percentages of the two forest systems in each of the agro-ecological zones. In each zone, the figure shows how many percentages of the farms in that zone chose each forest system. The boxes over the vertical bars indicate the ecosystem to which the vertical bars belong. There are six ecosystems: deserts, semi-arid, dry savannah, moist savannah, sub-humid, and humid forest. Each of these ecosystems, excluding the desert, is further classified by elevation: highland, mid-land, and lowland. In total, there are sixteen zones which are noted in the horizontal axis.

The figure unequivocally shows that the higher the precipitation of the zone, the higher the adoption percentage of the two forest systems. The adoption percentage is as high as 25% of all farms in the three sub-humid

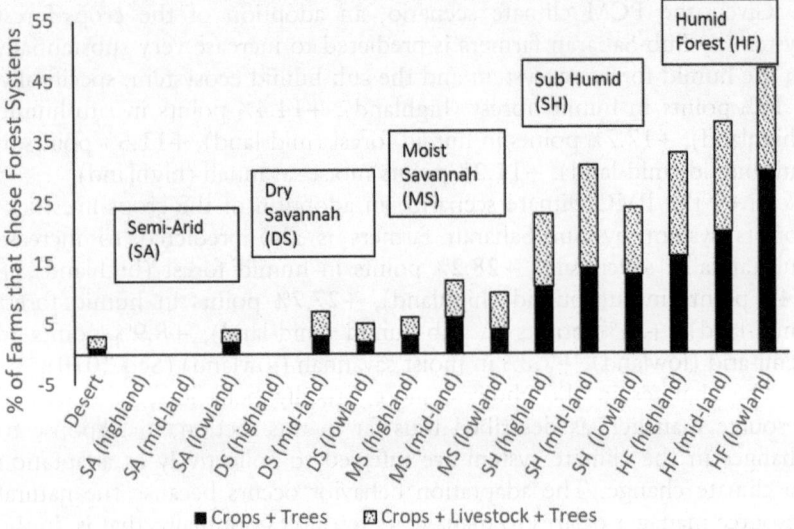

Fig. 3.5 Adoption of forest systems by Sub-Saharan Farmers

zones while it is as high as nearly 40% in the three humid forest zones. In the lowland humid forest zone, it is nearly 50%. The adoption percentage is drastically lower in dry savanna, semi-arid, and desert ecosystems.

The figure also indicates that the higher the temperature of the zone, the higher the adoption percentage of the two forest systems. To be more specific, farmers have chosen the two forest systems at higher percentages in the humid-forest and the sub-humid zones in lowlands and mid-elevations. These areas are highest temperature zones in Sub-Sahara. You can go back to Fig. 2.1 in Chapter 2 to review where these zones are located.

The information contained in Fig. 3.5 gives us a clue on how an individual resource manager in Sub-Sahara will adapt to changes in the climate system in the future by relying upon forest resources. Let's assume that one of the climate change predictions made by a highly respected climate prediction model will be realized. Specifically, the PCM (Parallel Climate Model) by the National Center for Atmospheric Research (NCAR) in the US predicts a 1.3 degree in Celsius increase in temperature as well as a 12% increase in precipitation in Africa by the year 2060 (Washington et al. 2000). The PCM is a hotter and wetter scenario for the future climate.

Given the PCM climate scenario, an adoption of the crops-forest system by Sub-Saharan farmers is predicted to increase very substantially in the humid forest ecosystem and the sub-humid ecosystem: specifically, +16% points in humid forest (highland), +14.4% points in sub-humid (highland), +17.7% points in humid forest (mid-land), +11.5% points in sub-humid (mid-land), +11.2% points moist savannah (highland).

Given the PMC climate scenario, an adoption of the crops-livestock-forests system by Sub-Saharan farmers is also predicted to increase substantially: specifically, +28.2% points in humid forest (highland), + 14% points in sub-humid (highland), +27.7% points in humid forest (mid-land), +21% points in sub-humid (mid-land), +8.9% points in semi-arid (lowland), +7.3% in moist savannah (lowland) (Seo 2010).

The changes in the choices, more broadly behaviors, of a natural resource manager, as described thus far in this section, in response to changes in the climate system are referred to collectively as adaptation to climate change. The adaptation behavior occurs because the natural resource manager desires to manage its resources optimally, that is, in the most profitable manner given the changing climate systems. An adaptation to a climatic change involves an increase in one system, which most often implies a decrease in another system. In the above analysis, because of the increases in the crops-forests system and the crops-livestock-forests system, the other systems, that is, the crops-only system and the crops-livestock system, will decline.

Another point to note is that an adaptation strategy to a climatic change scenario should be different from an adaptation strategy to another climatic change scenario. For example, if the climate system were to shift into a hotter and drier climate system by the middle of this century, the afore-described changes in the choices would not be realized. Owing to the more arid condition, the crops-livestock system, for instance, may increase across many ecosystems, which we already saw in Chapter 2.

A similar analysis was subsequently conducted for Latin American farms where the forest systems are even more salient (Seo 2012, 2016). Unlike in the African study, some Latin American farmers are observed to specialize in trees and forest products. Many rural households in Latin America can be classified as a forest-only enterprise. In addition, the crops-forests enterprise accounts for 12% of all farms in Brazil and 13% in Colombia, the two countries that own much of the Amazon rainforest.

A higher reliance on forest resources in Latin America be attributed to what the vast Amazon rainforest offers as well as the fact that rural households, besides timber and pulpwood sales, can collect edible and medicinal forest resources. The non-timber forest products by far outweigh timber products in the deep forest zones such as the Amazon rainforest (Peters et al. 1989). Rural households can consume the forest products by themselves but also sell them to the markets for income.

3.8 PLANTING TREES: FORESTS AND CLIMATE CHANGE INITIATIVES

The behavioral adaptation studies such as those described in the preceding section are beginning to be noticed and translated into a climate policy decision of many countries. Put differently, country policy-makers are increasingly recognizing their forests as a critical component of their climate policy designs. Let's consider some of the recent initiatives in this regard.

According to the United Nations Environment Programme, Ethiopia planted 350 million trees in 12 hours on July 2019 (UNEP 2019). The country's forest land has declined from 30% of the country at the beginning of the twentieth century to less than 4% today. In addition to being a forest coverage reclamation project, this nation-wide tree planting event was organized as a climate action. Through the Warsaw Framework at the United Nations, Ethiopia can ask for financial payments from the UN for the increased forest coverage and forest carbon stocks (UNFCCC 2016).

The Central African Forest Initiative (CAFI) was developed to preserve the forests in the region, where the Congo River forests are located (Fig. 3.1), as a regional policy action to fight climate change, reduce poverty, and for sustainable development of the region (CAFI 2020). The CAFI partners are Central African Republic, the Democratic Republic of the Congo, Cameroon, the Republic of Congo, Equatorial Guinea, and Gabon. The United Nations Development Programme (UNDP) administers the CAFI Fund. As we saw in this chapter, Central African countries already rely heavily on rich forest resources (Fig. 3.5).

On August 2019, Indians reported that the State of Uttar Pradesh, the most populous State of the country, planted 220 million trees in a single day (USA Today 2019). This was a part of the government's efforts to combat climate change and one million Indians participated in the planting in about 60,000 villages across the State. India has promised as part of the national commitment to the Paris Agreement to keep one-third of the country's land area under tree cover (UNFCCC 2015).

Finally, the Republican Party in the United States House of Representatives announced a three-pronged approach in which the first is encapsulated by the Trillion Trees Act. The legislation would look to increase the number of trees drastically in the US to capture carbon emissions and the credits may be given to the captured emissions (Fox News 2020). The Trillion Trees Vision, a joint venture by three world-renowned conservation agencies, plans to plant one trillion trees by 2050 as a strategy to combat climate change by connecting donors to forest ventures. President Trump announced that the US will join this effort (*The Hill* 2020).

3.9   Chapter Highlights

- This chapter tells the story of the Amazon rainforest and the Congo River rainforests, the two largest forests in the world, with their places in climate change.
- The forest is a central figure in the global carbon cycle: It absorbs carbon dioxide through photosynthesis and releases it through respiration.
- For the rural poor, forest resources are an important asset as a source of income and livelihoods in the Amazon rainforest. The value of non-timber forest products far outweighs the value of timber sales.
- The impacts of climate change on forest growth are researched via locality-based field experiments, from which changes in the forests are modeled at the national level as well as at the global level.
- A series of economic models finds that natural resource managers will adapt to climate change by increasing adoption of forestry and forest activities under a hotter and wetter climate condition, especially in sub-humid and humid forest ecosystems.
- The world communities are increasingly adopting a forest initiative as a strategy to fight global warming, e.g., a trillion trees vision.

REFERENCES

Ainsworth, E.A., and S.P. Long. 2005. What Have We Learned from 15 Years of Free-Air CO_2 Enrichment (FACE)? A Meta-Analysis of the Responses of Photosynthesis, Canopy Properties and Plant Production to Rising CO_2. *New Phytology* 165: 351–72.

Central African Forest Initiative (CAFI). 2020. How We Work. CAFI. Accessed from https://www.cafi.org/content/cafi/en/home/our-work/governance.html.

Ciais, P., C. Sabine, G. Bala, L. Bopp, V. Brovkin, J. Canadell, A. Chhabra, R. DeFries, J. Galloway, M. Heimann, C. Jones, C. Le Quéré, R.B. Myneni, S. Piao, and P. Thornton. 2013. Carbon and Other Biogeochemical Cycles. In *Climate Change 2013: The Physical Science Basis.* Cambridge: Cambridge University Pres.

Crowther, Thomas W., Henry B. Glick, Kristofer R. Covey, C. Bettigole, D.S. Maynard, S.M. Thomas, J.R. Smith, et al. 2015. Mapping Tree Density at a Global Scale. *Nature* 525 (7568): 201–5.

Faustmann, Martin. 1849. On the Determination of the Value Which Forest Land and Immature Stands Possess for Forestry [English edition edited by M. Gane, Oxford Institute Paper 42, 1968, entitled "Martin Faustmann and the Evolution of Discounted Cash Flow"].

Fox News. 2020. *GOP Looks to Counter Green New Deal with Three-pronged Climate Change Plan: Report.* New York: Fox News. Published on January 21, 2020.

Giacometti, D.C. 1990. Estratégias de coleta e conservação de germoplasma hortícola da América tropical. Proc. Simpósio Latinoamericano Sobre Recursos Genéticos de Espécies Hortícolas. Fundação Cargill, 91–110.

Harris, Nancy L., Sandra Brown, Stephen C. Hagen, Sassan S. Saatchi, Silvia Petrova, William Salas, Matthew C. Hansen, Peter V. Potapov, and Alexander Lotsch. 2012. Baseline Map of Carbon Emissions from Deforestation in Tropical Regions. *Science* 336 (6088): 1573–76.

Houghton, R.A. 2008. Carbon Flux to the Atmosphere From Land-Use Changes: 1850–2005. In *Trends: A Compendium of Data on Global Change.* Oak Ridge, TN: Carbon Dioxide Information Analysis Center, Oak Ridge National Laboratory, US Department of Energy.

Houghton, R.A., J.I. House, J. Pongratz, G.R. van der Werf, R.S. DeFries, M.C. Hansen, C. Le Quéré, and N. Ramankutty. 2012. Carbon Emissions from Land Use and Land-Cover Change. *Biogeosciences* 9 (12): 5125–42.

Joyce, Linda A., John Amber, Steve McNulty, Virginia Dale, Andrew Hansen, Lloyd Irland, Ron Neilson, and Kenneth Skog. 2000. Potential Consequences of Climate Variability and Change for the Forests of the United States. In *National Assessment Synthesis Team Climate Change Impacts on the*

United States: The Potential Consequences of Climate Variability and Change. Cambridge: Cambridge University Press.

Matthews, Elaine. 1983. Global Vegetation and Land Use: New High-Resolution Data Bases for Climate Studies. *Journal of Climate and Applied Meteorology* 22 (3): 474–87.

Mendelsohn, Robert. 2000. Efficient Adaptation to Climate Change. *Climatic Change* 45: 583–600.

Peters, Charles M., Alwyn H. Gentry, and Robert O. Mendelsohn. 1989. Valuation of an Amazonian Rainforest. *Nature* 339 (6227): 655–56.

Schlesinger, William H. 1997. *Biogeochemistry: An Analysis of Global Change*, 2nd ed. San Diego, CA: Academic Press.

Seo, S. Niggol. 2006. Modeling Farmer Responses to Climate Change: Measuring Climate Change Impacts and Adaptations in Livestock Management in Africa. PhD dissertation, Yale University, New Haven, CT.

Seo, S. Niggol. 2020. *The Economics of Globally Shared and Public Goods.* Amsterdam, NL: Academic Press.

Seo, S. Niggol. 2010. Managing Forests, Livestock, and Crops under Global Warming: A Micro-Econometric Analysis of Land Use Changes in Africa. *Australian Journal of Agricultural and Resource Economics* 54 (2): 239–58.

Seo, S. Niggol. 2012. Adaptation Behaviours Across Ecosystems Under Global Warming: A Spatial Micro-Econometric Model of the Rural Economy in South America. *Papers in Regional Science* 91 (4): 849–71.

Seo, S. Niggol. 2016. The Micro-Behavioral Framework for Estimating Total Damage of Global Warming on Natural Resource Enterprises with Full Adaptations. *Journal of Agricultural, Biological, and Environmental Statistics* 21 (2): 328–47.

Sohngen, Brent, and Robert Mendelsohn. 1998. Valuing the Impact of Large-scale Ecological Change in a Market: The Effect of Climate Change on US Timber. *American Economic Review* 88: 686–710.

ter Steege, H., N.C.A. Pitman, D. Sabatier, C. Baraloto, R.P. Salomao, J.E. Guevara, O.L. Phillips, et al. 2013. Hyperdominance in the Amazonian Tree Flora. *Science* 342 (6156): 1243092.

The Hill. 2020. *Trump Announces the US Will Join 1 Trillion Tree Initiative.* Washington, DC: The Hill. Published on January 21, 2020.

United Nations Environment Programme (UNEP). 2019. *Ethiopia Plants over 350 Million Trees in a Day, Setting New World Record.* Nairobi, KE: UNEP. Published on August 2, 2019.

United Nations Framework Convention on Climate Change (UNFCCC). 2015. *The Paris Agreement.* New York: UNFCCC.

United Nations Framework Convention on Climate Change (UNFCCC). 2016. *Key Decisions Relevant for Reducing Emissions from Deforestation and Forest Degradation in Developing Countries (REDD+).* New York, NY: UNFCCC.

USA Today. 2019. *Indians Plant 220 Million Trees in Single Day to Combat Climate Change*. Published on August 12, 2019.

Vedeld, Paul, Arild Angelsen, Espen Sjaastad, and Gertrude K. Berg. 2004. Counting on the Environment: Forest Incomes and the Rural Poor. Environmental Economics Series, Paper #98, The World Bank, Washington, DC.

Washington, W.M., J.W. Weatherly, G.A. Meehl, A.J. Semtner Jr., T.W. Bettge, A.P. Craig, W.G. Strand Jr., et al. 2000. Parallel Climate Model (PCM) Control and Transient Simulations. *Climate Dynamics* 16 (10–11): 755–74.

World Resources Institute (WRI). 2005. *World Resources 2005: The Wealth of the Poor: Managing Ecosystems to Fight Poverty*. Washington, DC: WRI.

CHAPTER 4

Indian Monsoon: A Tale of Indian Water Buffaloes, Goats, and High-Yield Rice

4.1 INDIAN MONSOON

The global climate system is a totality of the climate systems that are varied from one region to another. The Indian monsoon system is one of such regional climate systems, which is also one of the most feared ones. It rivals to the minds of many observers the harsh climate system in the Sahel where the climate system is hot, arid, and whimsical.

The present author grew up in an East Asian village which belongs to the temperate climate zone with four distinct seasons: Spring, Summer, Fall, and Winter. This climate system is similar to that of Western Europe around Paris and that of the US around Washington, DC.

What distinguishes the Indian monsoon system from the others and makes it feared by people? The system has two seasons: monsoon and off-monsoon. The system is hazardous because of a heavy rainfall during the monsoon season. The monthly monsoon rainfall may reach even 2,000 mm in some regions of India. For your reference, the average monthly precipitation during the wettest month of Washington DC is about 94 mm (IITM 2012; World Bank 2019).

The heavy rainfall during the rainy season of India which lasts about four months is dangerous, which is not difficult for anyone to understand. The heavy rains will sweep away crop fields, roads, bridges, shacks, and fragile houses. The number of dead or displaced people in India during

© The Author(s), under exclusive license to Springer Nature Switzerland AG 2021
S. N. Seo, *Climate Change and Economics*,
https://doi.org/10.1007/978-3-030-66680-4_4

the monsoon season is astounding, reaching often millions. As for farm-lands, soils will turn alluvial, floury, and dusty due to the heavy monsoon rains.

The second characteristic of the monsoon system is a starkly low precipitation during the off-monsoon season. The average monthly precipitation for December in India is only 10 mm per month, which means that many regions will experience nearly no rainfall at all (World Bank 2019). This is another very different water supply problem posed by the monsoon climate system during the off-monsoon. Indians rely on the water stored in community reservoirs and ponds during the monsoon season for agricultural and other residential activities for the off-monsoon season.

The third and perhaps most frightening feature of the monsoon climate system is a sudden shift from extremely high rainfall during the monsoon to extremely low rainfall during the off-monsoon, and vice versa. The sudden shifts back and forth in the weather will constrain economic activities severely in the monsoon climate region because the types of economic activities must be quite distinct in the two seasons. To give you an example, a rice farming during the non-monsoon season cannot be continued through the monsoon season due to heavy rain pours and flooding.

For anyone who is keen on global climate changes, the Indian monsoon should be a fascinating phenomenon to look into and make one wonder. What causes the Indian monsoon, so unique a regional climate system? How have Indians and their neighbors coped with the monsoon system through their own sensible choices and ways of life? Will the future global warming affect the monsoon system in India and, if so, will it make the monsoon climate far worse? What are the rational choices that Indians can make faced with the potential shifts in the monsoon climate regime caused by global climate changes?

This chapter is devoted to elucidating the Indian monsoon climate system, from which their unique ways of adapting to the challenges by Indian people will be highlighted. The exposition in this chapter can be applied, to some extent, to the Indian neighbors such as Bangladesh, Myanmar, and Sri Lanka which surround the Bay of Bengal. In partic-ular, I will come back to Bangladesh in this book at Chapter 5 with a focus on the tropical cyclones which have wreaked for so long havoc on the coastal communities of Bangladesh, with many of them causing over a hundred thousand human deaths.

4.2 Indian Monsoon and Climate Change

For many millenniums in the past since the Indian civilization emerged in the Indian sub-continent, the Indian monsoon climate system may have underdone abrupt shifts multiple times. During the past century in which Planet Earth has been gradually warming, scientists observed on the contrary a weakening trend of the monsoon rainfall in India, which they attributed to an increase in the release of black carbon and sulfate aerosols. At the same time, they reported an increase in extreme rainfall events over central India and other regions (Meehl and Hu 2006; Goswami et al. 2006).

A suite of climate prediction models from the CMIP5 (Climate Model Intercomparison Project) predicts on average an increase in mean precipitation in India over the twenty-first century. Further, it predicts an increase in interannual variability in precipitation as well as an increase in the number of extreme rainfall events. Further, the duration of the monsoon season is predicted to increase because of an earlier onset coupled with a later retreat. The prediction of a precipitation increase is attributed to the increased moisture flux from the ocean to land (Turner and Annamalai 2012; Christensen et al. 2013).

There remains, however, significant uncertainty over the predictions of the changes in the monsoon system caused by climate change. In particular, in a more realistic model which simulates the ENSO (El Nino Southern Oscillation)-monsoon relationship, it remains inconclusive on whether or not the number of extreme monsoon seasons will increase (Christensen et al. 2013).

To understand the monsoon climate system of India, we would need to define the monsoon season and the non-monsoon season first. Let's define the monsoon season by four summer months: June, July, August, and September. Non-monsoon season falls on all the other months then, of which the Indian data shows that the driest months are December, January, and February. As the first step, we can define average monthly precipitation for each season.

What indicators would capture the characteristics of the monsoon climate the best? The first indicator is the ratio of the monsoon season monthly precipitation over the non-monsoon season monthly precipitation. Let's call this ratio the monsoon precipitation ratio normal (MPRN). Note that we need to define the monthly precipitation as a long-term average, say, a 40-year average precipitation, e.g., from 1971 to 2010.

The MPRN would capture the intensity of a monsoon for each region (Seo 2016).

As I noted before, the monsoon season precipitation varies from one year to another year quite substantially, also from one decade to another decade. To capture this variability, we can construct the monsoon variability index (MVI) as the standard deviation of the MPR across the aforementioned 40 years. To be more scientific, we use the coefficient of variation (CV) in monthly precipitation which is the size of standard deviation independent of the mean precipitation of each region. The MVI would identify the monsoon climate's degree of variability across the 40-year period (Seo 2016, 2019a).

In Fig. 4.1, we can verify the unique characteristics of the Indian monsoon system. First, the precipitation ratio normal (the right-side vertical axis), which is the ratio of monsoon precipitation normal over non-monsoon precipitation normal, shows that many regions of India have very intense rainfall during the monsoon. Five States have the MPRN greater than 75 while many other states have the MPRN of about 20. This means that, on average, the monsoon season precipitation normal is 20 times the non-monsoon season precipitation normal for these States.

The MVI (the left-side vertical axis) also indicates that the monsoon season precipitation shifts suddenly from one year to another. The MVI

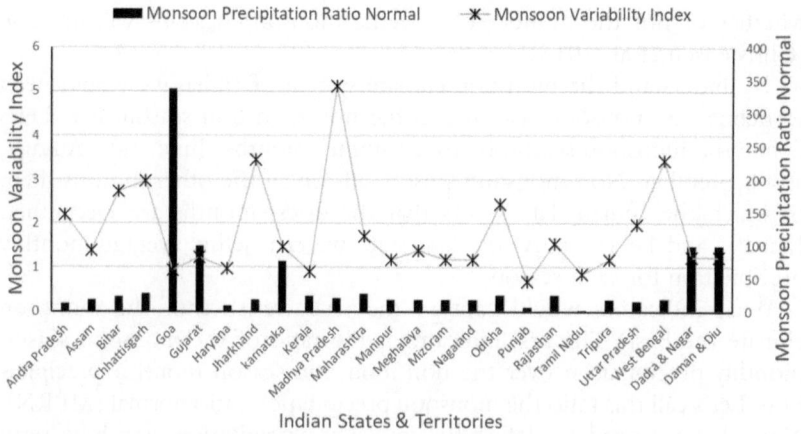

Fig. 4.1   Indian monsoon system

for Madhya Pradesh, Jharkhand, West Bengal, Chhattisgarh, Bihar, and Odisha is over 2.5 for each of these States. This means that on average the monsoon non-monsoon precipitation ratio shifts from one year to another by 250%. Additionally, the States of Andhra Pradesh and Uttar Pradesh have the MVI greater than 2. These States with a high MVI are the most heavily affected states by the monsoon climate system.

4.3 Monsoon, Indian Farming, and Poverty

In spite of the adversity posed by the monsoon climate system, India is gifted with vast lands and natural resources. In particular, India rivals only China in terms of the size of its population, which stands at about 1.35 billion as of April 2020 (World Bank 2020). Can India feed its gigantic population, despite the climate and weather adversity? If it can, how? This has long been a high-priority policy question for India, which has at the same time intrigued international observers and activists.

On top of the colossal population size, the country once suffered from a high poverty rate among its citizens. As illustrated in Fig. 4.2, the number of its citizens below the poverty line was as high as 60% of its total population during the 1970s. The poverty line, for clarity, is defined

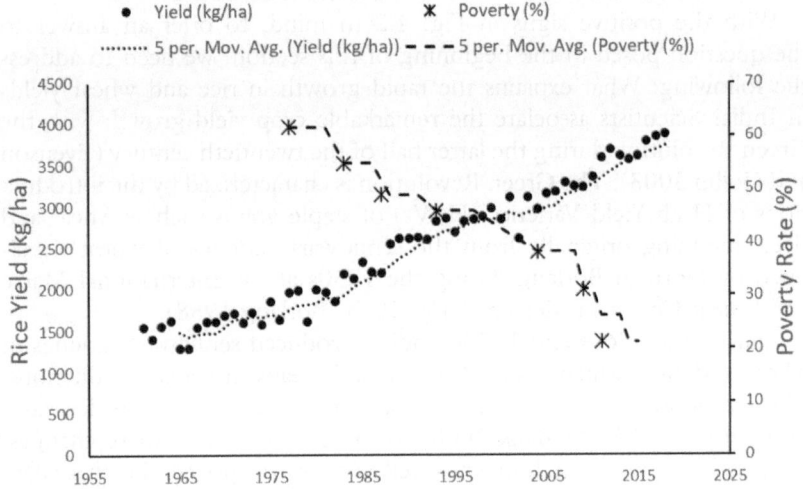

Fig. 4.2 Rice yield versus poverty rate in India since the 1960s

as the income of US$1.9 per day. At that time, the poverty rate in India was as high as those found in Sub-Saharan countries (World Bank 2020). As the figure reveals, the poverty rate has gradually declined since the 1970s to about 50% in the 1980s, to about 40% in the 1990s, to about 30% in the 2000s, and to about 20% in the 2010s.

The figure indicates that India has met the challenge of feeding its citizens, although the poverty rate is still very high compared with those in developed countries. The success is often attributed, as is indicated in Fig. 4.2, to the success of crop production in the country, especially the crop productivity growth up to the 2010s (Ravallion and Chen 2004; World Bank 2008). That is, the growth in the yields of staple grains in the country since the 1970s has been significant, which has played a key role in reducing the number of the poor in the country.

In Fig. 4.2, the history of rice yield since the 1960s is overlaid against the history of poverty rate over much of this time period in India. The figure shows the rice yield has increased almost linearly from 1.5 tons per hectare of land in the 1960s to 4 tons per hectare in the 2010s. Per unit land, Indian farmers are now producing paddy rice 250% of what was produced during the 1960s. The dramatic increase in rice yield shown in the figure is also found in other major cereals such as wheat (FAO 2020). The dramatic increase in rice yield shown in the figure is also found in other South Asian countries such as Thailand (Seo 2019a).

With the positive signs in Fig. 4.2 in mind, to offer an answer to the question posed in the beginning of this section, we need to address the following: What explains the rapid growth in rice and wheat yields in India? Scientists associate the remarkable crop yield growth with the Green Revolution during the latter half of the twentieth century (Evenson and Gollin 2003). The Green Revolution is characterized by the introductions of High Yield Varieties (HYVs) of staple grains such as wheat and rice, stemming originally from the semidwarf varieties of wheat developed by Norman Borlaug during the 1950s at the International Maize and Wheat Center in Mexico (CIMMYT) (Borlaug 1958).

During the 1960s and 1970s, India introduced semidwarf varieties of wheat and rice, which resulted in dramatic leaps in cereal productions. This was followed by an increase in farmer's income as well as decreases in grain prices (World Bank 2008). As a consequence of these changes, the number of people in hunger as well as below the poverty line has fallen steeply in India. I want to note that sometimes rice and wheat in India are planted in sequence through no-till agriculture (World Bank 2008).

This food revolution occurred not only in India but also in many South Asian, East Asian, and Latin American countries (Seo 2019a). The Green Revolution, however, did not materialize in Sub-Saharan Africa, which is one aspect of the revolution that I want to emphasize before we leave this section because of its relevance to many chapters in this book that look into the continent including Chapter 2 on farm animals and Chapter 3 on rainforests.

The introductions of the HYVs of wheat and rice did not result in higher yields of these cereals in Sub-Sahara, which is blamed on many factors such as climate, social, economic, political conditions. As shown in Fig. 4.3 for Mali, a Sahelian country, the yields of wheat and rice remained stagnant at around 1 ton per ha through the end of the twentieth century (World Bank 2009). The yields of both grains do show an increasing trend during the twenty-first century. The rice yield at 2014 stood at about 3.5 tons per ha while the wheat yield at about 4 tons per hectare of land.

Although the trends are encouraging, it is early to say that the Green Revolution is succeeding in Mali and other Sub-Saharan countries. The increase in cereal yields during the first 15 years of the twenty-first century is likely to be attributable to the unique rainfall regime in Sahel which has brought much higher rainfall during this time period (JISOA 2020). The

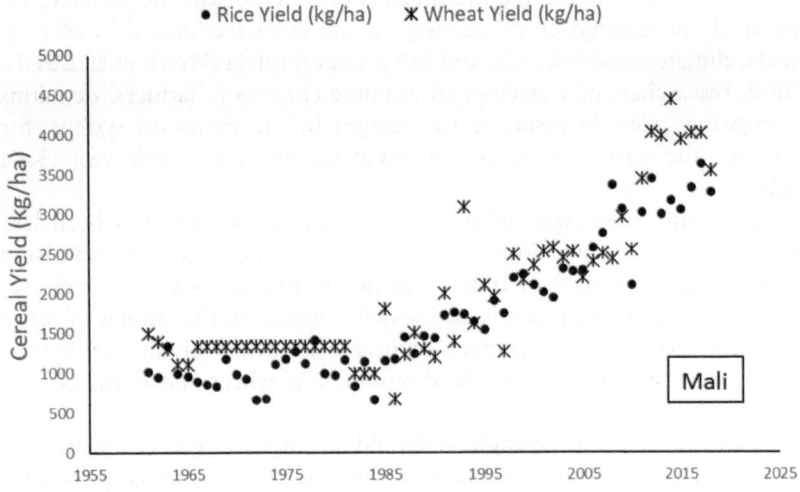

Fig. 4.3 Rice and wheat yields in Mali, Sub-Sahara since the 1960s

rainfall regime swings every two decades or so, which is caused by changes in the ocean circulation called Atlantic Multidecadal Oscillation. After the high rainfall period, the region will certainly swing back again to a low rainfall period. It has yet to be seen whether Mali farmers could overcome the dry period to sustain the cereal yields achieved during the twenty-first century.

4.4   Indian Agriculture and Climate Change

Given the success of Indian agriculture on two fronts of both crop yields and the poverty rate, it is quite pertinent to ask whether climate change will affect the Indian agriculture in a harmful way or even reverse the trends in the decades to come. Further, any answer to this question would invariably depend on changes in the monsoon climate system.

A prediction of the impacts of climate change on Indian agriculture is not at all easy, even if we assume that a specific climate change scenario would unfold. First of all, researchers can attempt to measure the impacts of the climate change scenario on a selected crop or another across the range of crops (Aggarwal and Mall 2002). This is commonly done by the field experiments in which the impact of an increase in the level of carbon dioxide on the yield of a chosen crop. Second, researchers alternatively attempt to examine the impacts of different climate variables on the yields of selected crops, relying on the cross-sectional data of crop yields, climate variables, soils, and other relevant data (Welch et al. 2010). Third, researchers may attempt to examine changes in farmers' decisions in growing crops in response to changes in the monsoon system, for example, the start date of the monsoon season in a specific year (Kala 2015).

Across the three types of studies, a focus of attention has been laid heavily on rice farming. From one angle, it would be quite understandable considering the gravity of Indica rice in the Indian society and culture. This priority crop approach is also widely adopted in the studies of other countries. For example, climate economic modelers in the US gave a priority to the analysis of staple crops such as wheat and corn, besides soybeans.

From a more wholistic angle, it should be emphasized that Indian agriculture is far greater than and far more diverse than rice farming. Besides the two staple cereals, Indian agriculture is famed for a great diversity of pulses (including lentil beans), potatoes, and oilseeds, among other things. Specialty crops of India include tea (including Darjeeling tea), jute, rubber, coffee, among other things. Other than the different types of crops, India is also famed for milk production from cows, buffaloes, and goats. As of 2005, the Indian Dairy Cooperatives had over 12 million members (World Bank 1998; Seo et al. 2005).

The field experiments of crop growth indicated that carbon dioxide increase in the atmosphere would be beneficial to crop growth. However, the increases in temperature would be very damaging to Indian crop agriculture, especially the increase in daily minimum temperature (Welch et al. 2010). The authors reported, through the South Asian study of rice yield, that about a half of a degree increase in temperature (in Celsius) in the past decades resulted in a decrease in rice yield by as much as 10–25% in many prime crop-growing regions of South Asia. What this implies is that a 3 or 4 degree increase in temperature (in Celsius) by the end of this century may result in a severe loss in yields of Indian primary cereals (Aggarwal and Mall 2002).

4.5   WATER BUFFALOES AND GOATS

The studies of crop yield changes under climate change in India, explained in the previous section, however, do not shed much light on the influences of the Indian monsoon system. Neither do they make clear the Indian ways of living that were invented to cope with the whims of the unique monsoon system. One of such creative strategies would be an adjustment of farming activities in response to the onset date of a yearly monsoon season (Kala 2015).

The severe monsoon climate system in India as highlighted in Fig. 4.1 makes us wonder fundamentally how Indian people have coped with the climate and weather reality of the country for so long. Further, we are led to wonder what economic activities rural areas of India are engaged with during the intense monsoon seasons.

If you are acquainted with the Indian classical texts such as the Vedas or the collection of Buddha's teachings, you would imagine that monks and laypersons would go into a long retreat during the rainy season. Most of the crop agriculture, if not all, would be certainly made impossible during the monsoon. The trees on the fields and mountains owned by farmers would continue to grow but fruits would fall and rot due to heavy rains. Many farms would resort to non-farming activities such as textiles, baskets, etc. Further, a large fraction of farmworkers would migrate to other States in search of a manufacturing or service job.

In the monsoon season, Indians can turn to fiber productions such as silk, jute, and wool, which forms the basis of the textile industry of India. They may grow silkworms and mulberries through the season to produce silks. The silks can be sold in the market for income. Similarly, they may produce cashmere wool out of cashmere goats.

Another resort of Indians is animal husbandry. Milk and milk products are a critical component of food security in India, especially for nutritional security. In 2018, India produced about 186 million tons of milk per year. India is the largest producer of milk worldwide: About 22% of the world's milk production comes from India (World Bank 1998). Over 80 million households from about total 147 million households depend on dairy production for their livelihood. There are two farm animals quite noteworthy in this regard, water buffaloes and goats.

The water buffalo is regarded as a sacred animal in the Indian classical religion. More pertinently, it is regarded as your mother in your previous lives in the cycle of rebirths. They are treated with worship, as such, no meat is obtained from the water buffaloes, alive or dead. However, the water buffaloes are a primary source of energy and nutrition for Indian families through milk production.

The goat seems to be treated very differently. The present author felt during the stay in India that the goats are kicked around by Indian people, literally. Locals refer to the goats as an ATM (Automated Teller Machine), meaning that they can make easy and quick cash when needed by selling goats in the market or selling goat meats and milk. The cashmere wool, one of the signature exports of India for so long, is made out of goat wool.

When it comes to the impacts of climate change, the goat has been reported that the farm animal is preferred and raised more often in a hot and high rainfall zone of Sub-Saharan Africa (Seo and Mendelsohn 2008). Further, as we discussed in Chapter 2, the mixed crops-livestock system of agriculture is favored by Sub-Saharan farmers when the climate system turns to a high variability of rainfall interannually as well as intra-annually (Seo 2012). We can therefore imagine that the goats are also widely adopted in the high temperature, high precipitation, and high variability condition that characterizes the Indian monsoon system.

The graph in Fig. 4.4 indicates that the hypothesis may actually be true. This was the finding that was first presented in the paper that examined

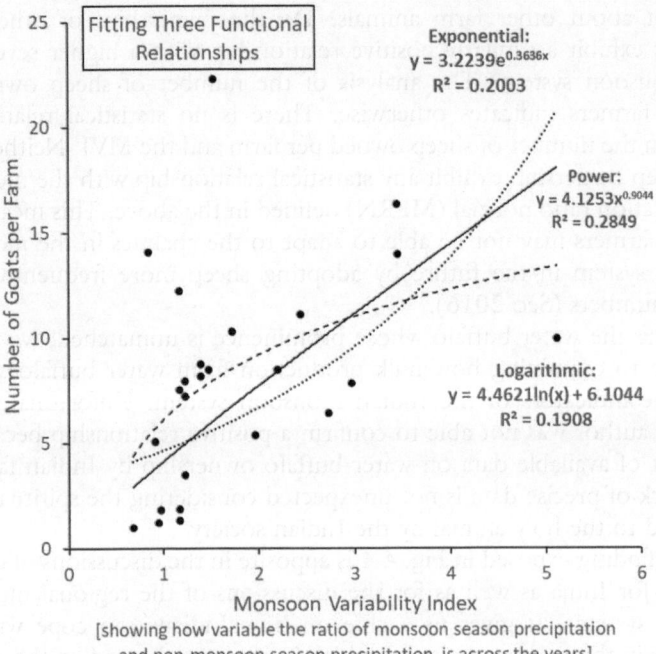

Fig. 4.4 Goats and the Indian monsoon

how Indian farmers have adapted to the regional monsoon system. The figure shows that there exists a positive relationship between the number of goats per farm and the monsoon variability index (MVI), defined above in Fig. 4.1, in a statistically significant manner (Seo 2016).

Specifically, the figure indicates that the higher the MVI, the larger the number of goats owned per farm. To make the figure clearer, I used the number of goats per 10 rural farms. Three statistical relationships are fit to the original data and overlaid in the figure: exponential, logarithmic, and power law. The three regressions indicate that the relationship is positive while the best fit is found in the power-law function. From the theoretical perspective, we should prefer the logarithmic function or the power-law function to the exponential function because the number of goats per farm cannot increase infinitely.

What about other farm animals? Do the ownerships of other farm animals exhibit a similarly positive relationship with a higher severity of the monsoon system? The analysis of the number of sheep owned by Indian farmers indicates otherwise. There is no statistical relationship between the number of sheep owned per farm and the MVI. Neither does the sheep ownership exhibit any statistical relationship with the monsoon precipitation ratio normal (MPRN) defined in the above. This means that Indian farmers may not be able to adapt to the changes in the monsoon climate system in the future by adopting sheep more frequently or in larger numbers (Seo 2016).

As per the water buffalo whose prominence is unmatched, we should be keen to examining how milk production from water buffaloes varies with the indicators of the Indian monsoon system. Unfortunately, the present author was not able to confirm a positive relationship because of the lack of available data on water buffalo ownership by Indian families. This lack of precise data is not unexpected considering the spiritual value attached to the holy animal by the Indian society.

The finding exposed in Fig. 4.4 is apposite in the discussions of climate change for India as well as for the discussions of the regional monsoon climate systems. It gives us a clue on how Indians can cope with the changes in the monsoon system, which is often overlooked by the climate change experts whose attention is fixated on staple crops, say, rice and wheat (Hahn 1981). More specifically, although a strengthening of the regional monsoon system could be devastating to a certain group of grain farmers, it may benefit a certain group of livestock managers. The differential effects would force Indian farmers to decrease crop farming while simultaneously increasing animal husbandry.

From the scientific vantage point, it should be a high-priority research to uncover genetical factors that make goats and other farm animals more tolerant of or resilient to certain climatic conditions (Hahn et al. 2009; NRC 2015). What makes goats more resilient to the Indian monsoon system from the genetics viewpoint? The answer to this question, if successfully discovered, would be as monumental as the answer to the question that began the Green Revolution: that is, what makes wheat yield higher?

4.6   Livelihoods in the Indian Monsoon

The impacts of the Indian monsoon system on the Indian society are pervasive, therefore, cannot be satisfactorily explained by the impacts on agriculture and livestock management alone. The impacts are wide ranging even beyond the totality of the economic system. The limitations notwithstanding, the focus laid on food and nutritional security by the researchers in the past is also justifiable.

With a wholistic view, let me point out some of the areas of grave concern with regard to the changes in the monsoon. First, a heavy monsoon rainfall, more often than not, results in displacements of a large number of Indians annually, sometimes millions of them. This is attributable mainly to the fragility of the houses that people live in. A severe flooding would not fail to sweep away many types of houses, e.g., a thatched house, a mud house.

Second, a monsoon season also forces a large-scale migration of Indian people from one region to another in search of a seasonal employment and income during the season. At the end of the monsoon season, these migrants often come back to their villages to farm.

Third, besides the heavy rainfall and severe drought ascribable to the monsoon system, Indians live with a high temperature climate system, that is, a tropical climate. India hosts many of the hottest places on the Planet where summer temperature exceeds 50 degrees in Celsius. Even setting aside the hottest places, India is hot across the country as well as across the year. To endure the heat stress, you will find out handily that Indians prefer a mud house. The mud-house is a house built solely by red mud. Owing to the mud density, the solar radiation and high heat do not pass through the mud structure, making inside the house cool even during the summer season. You cannot expect this cooling effect from the houses built from cement or bricks, which are popular in temperate climate zones, which heat up the inside of the house. Neither can you expect this effect from the granite houses that are popular in the high-latitude climate zones, say, northern Europe.

Fourth, the monsoon system coupled with a high variability in the amount of rainfall both interannually and intra-annually, as measured by the MPRN or the MVI as defined above, makes it onerous to build the sewage system in urban and rural areas. A heavy flooding may overwhelm the sewage system. This resulted in the lack of a modern toilet system in many areas of India. This in turn resulted in not a few public health problems including vector-borne, say, mosquito-carrying, diseases such as Malaria.

I want to emphasize that the list of issues described in this section are not inherently a climate change-caused problem. These problems had exited long before and should be addressed by a sustained economic growth and subsequently higher income for the people. I hope I made a compelling case that when it comes to India, the economic growth should be a top priority of the nation, rather than a climate change policy (Sen 1990). You will also find that this recognition is also well reflected in the India's negotiation positions exhibited in the international climate policy negotiations, *e.g.*, Copenhagen Accord and Paris Agreement (UNFCCC 2015; Seo 2019b).

4.7   CHAPTER HIGHLIGHTS

- A monsoon system is one of the most salient regional climate systems, of which the Indian monsoon is famed for.
- The Indian monsoon system can be captured by the monsoon precipitation ratio and the monsoon precipitation variability index.
- Despite the monsoon, India was successful in feeding its 1.3 billion people through adopting the high-yielding varieties of rice and wheat through the Green Revolution.
- Climate change may harm the yields of these cereals in India seriously.
- Animal husbandry through water buffaloes and goats, which is an essential component for national food and nutritional security, may turn out to be a key strategy in the country's efforts to adapt to the Indian monsoon system.
- The effects of the monsoon are, however, far overreaching beyond agriculture, for which a sustained economic growth should remain a top priority despite the climate change concerns.

REFERENCES

Aggarwal, P.K., and R.K. Mall. 2002. Climate Change and Rice Yields in Diverse Agro-Environments of India. II. Effect of Uncertainties in Scenarios and Crop Models on Impact Assessment. *Climatic Change* 52: 331–43.

Borlaug, Norman E. 1958. The Impact of Agricultural Research on Mexican Wheat Production. *Transactions of the New York Academy of Science* 20: 278–95.

Christensen, J.H., K.K. Kumar, E. Aldrian, S. An, I.F.A. Cavalcanti, M. de Castro, W. Dong, P. Goswami, A. Hall, J.K. Kanyanga, A. Kitoh, J. Kossin, N. Lau, J. Renwick, D.B. Stephenson, S. Xie, and T. Zhou. 2013. Climate Phenomena and Their Relevance for Future Regional Climate Change. In *Climate Change 2013: The Physical Science Basis*. Cambridge: Cambridge University Press.

Evenson, Robert E., and Douglas Gollin. 2003. Assessing the Impact of the Green Revolution, 1960 to 2000. *Science* 300 (5620): 758–62.

Food and Agriculture Organization (FAO). 2020. *FAO STAT*. Rome, IT: FAO.

Goswami, B.N., V. Venugopal, D. Sengupta, M.S. Madhusoodanan, and P.K. Xavier. 2006. Increasing Trend of Extreme Rain Events Over India in a Warming Environment. *Science* 314 (5804): 1442–45.

Hahn, G. LeRoy. 1981. Housing and Management to Reduce Climatic Impacts on Livestock. *Journal of Animal Science* 52 (1): 175–86.

Hahn, G. LeRoy, J.B. Gaughan, T.L. Mader, and R.A. Eigenberg. 2009. Thermal Indices and Their Applications for Livestock Environments. In *Livestock Energetics and Thermal Environmental Management*, ed. J.A. DeShazer, Chapter 5, pp. 113–30. St. Joseph, MI: ASABE. Copyright 2009 American Society of Agricultural and Biological Engineers.

Indian Institute of Tropical Meteorology (IITM). 2012. *Homogeneous Indian Monthly Rainfall Data Sets & Indian Regional Monthly Surface Air Temperature Data Set*. Pune, India: IITM.

Joint Institute for the Study of Ocean and the Atmosphere (JISOA). 2020. *Sahel Precipitation Index*. Seattle, WA: JISOA, University of Washington.

Kala, Namrata. 2015. Ambiguity Aversion and Learning in a Changing World: The Potential Effects of Climate Change from Indian Agriculture. PhD dissertation, Graduate School of Arts and Sciences, Yale University, New Haven, CT.

Meehl, Gerald A., and Hu Aixue. 2006. Megadroughts in the Indian Monsoon Region and Southwest North America and a Mechanism for Associated Multidecadal Pacific Sea Surface Temperature Anomalies. *Journal of Climate* 19 (9): 1605–23.

National Research Council (NRC). 2015. *Critical Role of Animal Science Research in Food Security and Sustainability*. Washington, DC: The National Academies of Sciences, Engineering, and Medicine (NASEM).

Ravallion, Martin, and S. Chen. 2004. How Have the World's Poorest Fared Since the Early 1980s? *The World Bank Research Observer* 19 (2): 141–69.

Sen, Amartya. 1990. *Development as Freedom*. New York: Anchor Books.

Seo, S. Niggol. 2012. Decision Making Under Climate Risks: An Analysis of Sub-Saharan Farmers' Adaptation Behaviors. *Weather, Climate, and Society* 4 (4): 285–99.

Seo, S. Niggol. 2016. Untold Tales of Goats in Deadly Indian Monsoons: Adapt or Rain-Retreat under Global Warming? *Journal of Extreme Events* 03 (01): 1650001. https://doi.org/10.1142/s2345737616500019.

Seo, S. Niggol. 2019a. Will Farmers Fully Adapt to Monsoonal Climate Change through Technological Developments? An Analysis of Rice and Livestock Production in Thailand. *The Journal of Agricultural Science* 157 (2): 97–108.

Seo, S. Niggol. 2019b. *The Economics of Global Allocations of the Green Climate Fund: An Assessment from Four Scientific Traditions of Modeling Adaptation Strategies*. Cham: Springer Nature.

Seo, S. Niggol, and Robert Mendelsohn. 2008. Measuring Impacts and Adaptations to Climate Change: A Structural Ricardian Model of African Livestock Management. *Agricultural Economics* 38 (2): 151–65.

Seo, S. Niggol, Robert Mendelsohn, and Mohan Munasinghe. 2005. Climate Change and Agriculture in Sri Lanka: A Ricardian Valuation. *Environment and Development Economics* 10 (5): 581–96.

Turner, Andrew G., and H. Annamalai. 2012. Climate Change and the South Asian Summer Monsoon. *Nature Climate Change* 2 (8): 587–95.

United Nations Framework Convention on Climate Change (UNFCCC). 2015. The Paris Agreement. In *Conference of the Parties (COP) 21*. New York: UNFCCC.

Welch, Jarrod R., Jeffrey R. Vincent, Maximilian Auffhammer, Piedad F. Moya, Achim Dobermann, and David Dawe. 2010. Rice Yields in Tropical/Subtropical Asia Exhibit Large but Opposing Sensitivities to Minimum and Maximum Temperatures. *Proceedings of the National Academy of Sciences* 107 (33): 14562–67.

World Bank. 1998. India's Dairy Revolution. World Bank Operations Evaluation Department, #168. Washington, DC: The World Bank.

World Bank. 2008. *World Development Report 2008: Agriculture for Development*. Washington, DC: World Bank.

World Bank. 2009. *Awakening Africa's Sleeping Giant: Prospects for Commercial Agriculture in the Guinea Savannah Zone and Beyond*. Washington, DC: World Bank and FAO.

World Bank. 2019. *Climate Change Knowledge Portal for Development Practitioners and Policy Makers*. Washington, DC: The World Bank.

World Bank. 2020. *World Bank Development Indicators*. Washington, DC: The World Bank.

CHAPTER 5

A Refuge from Oceans and Hurricanes: A Story of Cyclone Shelters in Bangladesh Abutting the Bay of Bengal

5.1   THE WORLD OCEANS AND CLIMATE CHANGE

As far as the changes in the global climate system are concerned, the oceans of the Planet cannot be placed at sidelines. Simply, the oceans cover approximately 72% of the Earth's surface, far outweighing the coverage of the land ecosystems. The oceans are a habitat for numerous ocean species, in the range of several millions of marine species including coral reefs (UN 2017). Further, the world's oceans offer vast resources to the humanity, including food resources and petroleum.

The dependence of humanity on the oceans can be best explained by the number of people living on coastal zones and cities as well as the food resources that humanity obtain from the oceans. Over 600 million people, about 10% of the world's population, live in coastal areas located at less than 10 meters above the sea level. Fish from the oceans provide about 17% of the protein consumption by humans at the global level and over 50% in many developing countries. Fishing is a major source of income and food to the world's fishermen (UN 2017).

What do the oceans have to do with climate change? To begin with, Earth's temperature cannot be measured precisely without measuring them at the oceans. However, measuring temperatures at the oceans is by far more difficult than measuring temperatures in the land surfaces because weather stations cannot be placed over sea surfaces in sufficient numbers. Nowadays, numerous weather satellites observe the ocean surfaces to measure temperatures there.

At the core of the oceans' role in climate change lies the atmosphere ocean exchange of carbon dioxide, through which process the oceans play a pivotal role in setting the temperature of the Planet (Revelle and Suess 1957). The oceans are a reservoir of carbon dioxide from the atmosphere storing vast amount of carbon there. Annually, the oceans sink 80 GtC from the atmosphere and release 78 GtC into the atmosphere (Ciais et al. 2013). The oceans store over 40,000 GtC in them which is over an order of magnitude larger than the total storage by the land ecosystems.

The net ocean carbon uptake has been increasing to roughly 2.5 GtC or PgC (Petagrams of Carbon) per year at the time of this writing, which may continue to increase in the coming decades (Feely et al. 2020). The net uptake is occurring in the oceans at high latitudes in both hemispheres while the net release of carbon is occurring in the tropical oceans. The increase in the ocean carbon uptake could speed up ocean acidification.

Another route that the oceans can alter the Earth's temperature and climate system is through the atmosphere ocean heat exchange. The oceans are a reservoir of heat. Averaged over different layers of the oceans, the heat gain rate was 0.55–0.79 watts per square meter for the time period of 1993–2019. More than 90% of the warming that occurred on Earth during 1971 to 2010 occurred in the oceans (Johnson et al. 2020). An increase in the ocean heat content is contributing to numerous events such as sea level rises, melting of glaciers and ice sheets, ocean heat waves, and an increase in marine clouds, changes in fish catch and fish migrations, and changes in marine species.

The global sea level has risen from 1901 to 2010 at the rate of 1.7 mm per year. Some models predict a catastrophic rise in the sea level, as high as 5 meters or even higher, by the end of the century, which is powered by both thermal expansion of the oceans and the collapses of the West Antarctic and other ice sheets (Refer to Fig. 9.4 in Chapter 9 for the map of the West Antarctic) (Church et al. 2013).

Although invisible to bare eyes, the ocean current circulations in the world oceans constitute a critical component of the Earth system, in particular, the climate system. The great ocean conveyor belt, also called the Thermohaline Circulation, carries the cold water to the south all the way to the Antarctica and the warm water to the North Sea and the Arctic. A slowdown in the ocean current circulations may occur by way of changes in ocean temperature differentials, which is predicted to harm northern European countries (Broecker 1997).

Of all the ocean events, the climate and ocean connection that has received the most attention from both the public and researchers is tropical cyclones (Emanuel 2008). I will tell a story of tropical cyclones in this chapter, especially with a focus on where such events are deadliest, the Bay of Bengal. Bangladesh is a low-lying country in South Asia adjoining the Bay of Bengal. The country is particularly notorious for its high vulnerability to the tropical cyclones generated in the Bay of Bengal in the Indian Ocean. Some cyclones were reported to have killed more than a hundred thousand people each (Murty et al. 1986; Ali 1999).

The fear widely shared among the global citizens is that a warmer global ocean temperature caused by a Planetary warming may increase the number of intense tropical cyclones as well as the intensity of each cyclone, wreaking havoc on the particularly vulnerable nations such as Bangladesh (Seo and Bakkensen 2017).

5.2   Tropical Cyclones

Of all the oceanic climate events described in the preceding section, some events may actually be realized at a future date catastrophically for the globe. Of these ocean-generated potential catastrophes, tropical cyclones have received the most attention from the public owing to the familiarity of the world citizens with hurricane catastrophes, in addition to recent advances in the climate and hurricane research. Across the Planet, some 90 tropical cyclones are generated and some of them make landfalls each year, of which several cyclones turn out to be deadly in most years. As such, everyone knows well what a hurricane catastrophe looks like.

Hurricanes, tropical cyclones, and typhoons are the terminologies used interchangeably by scientists to describe a fast-moving storm generated in an ocean. In the North Atlantic and the Northeast Pacific Ocean, it is called a hurricane. In the North Indian Ocean and the Southern Hemisphere, it is called a tropical cyclone. In the Northwest Pacific Ocean, it is called a typhoon (McAdie et al. 2009).

What is a hurricane then? You know what it is, what it looks like, and most likely have experienced at least one strong hurricane in your life. It is not an easy job, though, to explain a hurricane scientifically. The simplest explanation is given by a feedback loop as follows. The flux of heat somewhere in the ocean generates surface winds in the ocean.

Surface winds then generate lower pressure there. The Lower pressure generates an increase in the heat flux. The increased heat flux then generates even stronger winds. The "heat engine" rotates counterclockwise (in the Northern Hemisphere) and moves erratically with high speeds. When the speeds of surface winds reach a certain threshold value, it is called a hurricane (Emanuel 2008).

If the maximum sustained winds of a storm exceed 17 meters per second (mps), it is called a tropical storm. When the maximum sustained winds exceed 33 mps, then it is called a tropical cyclone in the Pacific Ocean and a hurricane in the North Atlantic Ocean. For your reference, 33 mps corresponds to 119 km per hour, or 74 miles per hour (mph), or 64 nautical miles per hour.

The severity of a hurricane is often described by the Saffir-Simpson scale which is defined by the maximum sustained wind speeds (MSWS) (Saffir 1977). The MSWS is the maximum of 1-minute average surface wind speeds. The five categories of a hurricane are as follows:

Category 1 if MSWS $\geq$ 74 mph;
Category 2 if MSWS $\geq$ 96 mph;
Category 3 if MSWS $\geq$ 111 mph;
Category 4 if MSWS $\geq$ 130 mph;
Category 5 if MSWS $\geq$ 158 mph.

Other than the wind speeds, there are many other aspects of a hurricane that are pertinent in characterizing a hurricane, which makes hurricane science complicated. These aspects include minimum central pressure, places of origin, cyclone tracks, wind shear (i.e., the difference in wind speed along a vertical straight line of a storm), hurricane size, vorticity (i.e., a measure of rotation), duration, accompanying rainfall, storm surge height, landfall locations (McAdie et al. 2009). Many of these variables will be reintroduced in the upcoming sections at appropriate moments.

Before I leave this section, I need to emphasize and state explicitly the climate and hurricane connections. In the above description, you may have guessed, correctly so, that the heat flux and the heat engine may be dependent on the changes in the climate system which affect the sea surface temperature. In addition, scientists report that a hurricane cannot be generated when the sea surface temperature is below 26.5 degrees

(Celsius). Further, the sea surface temperature places a thermodynamic upper limit on the intensity of a hurricane (Knutson et al. 2010).

5.3 CYCLONE FATALITIES IN BANGLADESH IN THE INDIAN OCEAN

Tropical cyclones are feared by the public because of a large number of fatalities often caused by an intense tropical cyclone strike. In the US, Hurricane Katrina in 2005, which was the natural disaster which enraged the public unlike the others, flooded the city of New Orleans and killed 1833 people. Of all the coastal regions in the world, the countries that surround the Bay of Bengal (BOB) in the Indian Ocean have suffered the worst in terms of fatalities from tropical cyclones.

The Bay of Bengal (BOB) along with its neighborhood countries are drawn in Fig. 5.1 as part of the world oceans. The Bay of Bengal is a half-moon area in the Indian Ocean to the right of India and to the left of Thailand. The neighborhood countries, from the left to the right, are Sri Lanka, India, Bangladesh, Myanmar, and Thailand. The coastal areas in Bangladesh and India that abut the Bay of Bengal are low-lying lands, with much of them below 10 meters above the sea level, which makes the coastal communities particularly vulnerable (Dube et al. 2009).

The figure marks the eight world ocean regions used for classifying tropical cyclone regions, of which the Bay of Bengal belongs to Region V (NHC 2020):

Region I: Atlantic;
Region II: Eastern Pacific;
Region III: Central Pacific;
Region IV: Northwest Pacific;
Region V: North Indian Ocean;
Region VI: Southwest Indian Ocean;
Region VII: Southwest Pacific;
Region VIII: South Pacific.

In addition to being lowlands, the Bengal Delta in Bangladesh and the West Bengal State of India is the world's largest river delta through which the waters from the two world's largest rivers flow down to the Bay: Ganges River in India and Yarlung Tsangpo River in Tibet. Because

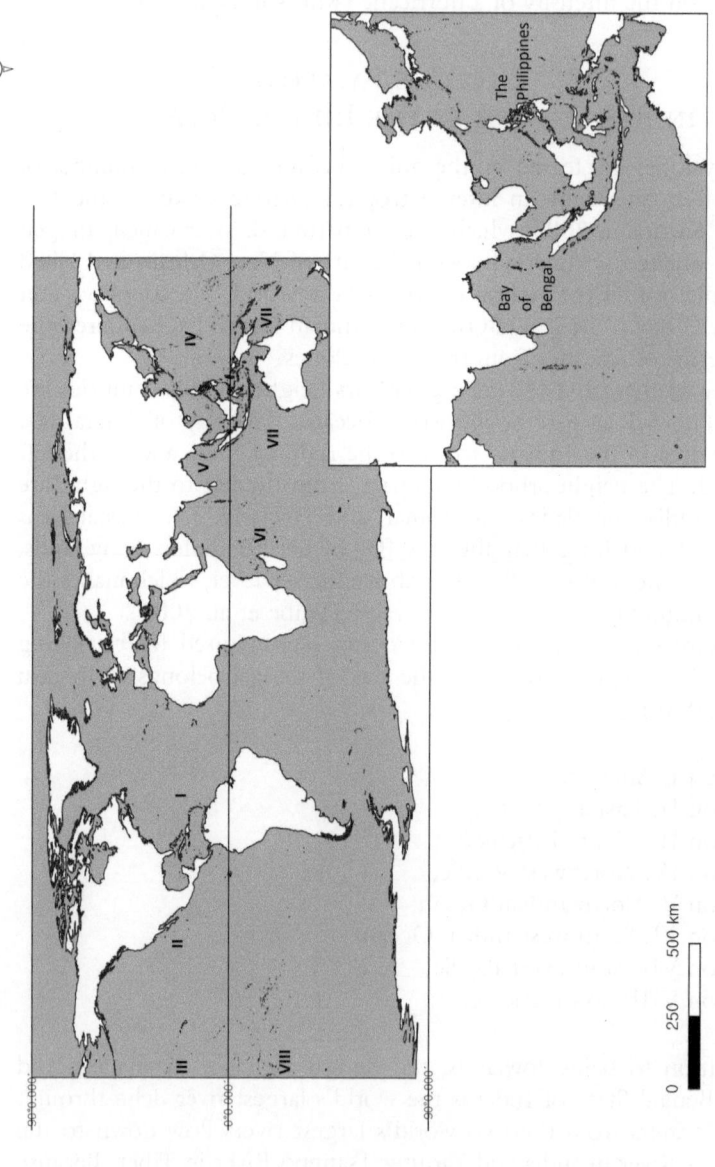

Fig. 5.1 World Ocean Regions and Bay of Bengal (*Note* I = Atlantic Ocean; II = Eastern Pacific; III = Central Pacific; IV = Northwest Pacific; V = North Indian Ocean; VI = Southwest Indian Ocean; VII = Southwest Pacific and Southeast Indian Ocean; VIII = South Pacific Ocean)

of the river delta, the coastal lands are very fertile for crop agriculture, as such home to a large number of people of about 400 million. The geo-economic conditions make the BOB countries even more vulnerable to tropical cyclones (Ali 1999).

The high risk faced by the BOB neighborhoods to tropical cyclones is confirmed by the historical data on human deaths. In Table 5.1, I summarize the deadliest tropical cyclones historically during the twentieth and twenty-first centuries in terms of the number of resultant fatalities, drawing from various data sources (Ali 1999; Bakkensen and Mendelsohn 2016; Cerveny et al. 2017). Of the 10 deadliest cyclones, 6 cyclones were generated from the Indian Ocean, all of them from the Bay of Bengal.

Table 5.1 Ten deadliest Tropical Cyclones during twentieth and twenty-first centuries

Rank	Year	The Ocean Region of Origin	Countries most affected	Cyclone name	Intensity: Saffir-Simpson Scale	Number of Fatalities
1	1970	Indian Ocean	Bangladesh	Cyclone Bhola	Category 3	300,000 to 500,000
2	1975	Northwestern Pacific	China, Philippines	Typhoon Nina	Category 4	229,000
3	1991	Indian Ocean	Bangladesh	Cyclone Gorky	Category 5	138,866
4	2005	Indian Ocean	Myanmar	Cyclone Nargis	Category 4	138,366
5	1977	Indian Ocean	India	Andhra Pradesh Cyclone	Category 4	Up to 14,204
6	1963	Indian Ocean	Eastern Pakistan [Bangladesh at present]	Severe Cyclonic Storm Three	Category 5	11,520
7	1998	North Atlantic	Honduras, Nicaragua	Hurricane Mitch	Category 3	11,374
8	1999	Indian Ocean	India	Odisha Cyclone	Category 5	At least 10,000
9	2013	Northwestern Pacific	Philippines	Typhoon Haiyan [Super Typhoon Yolanda]	Category 5 [315 km/h]	6,329
10	2005	North Atlantic	USA	Hurricane Katrina	Category 3	1,833

Although not shown in the table, historical records indicate that there were half a dozen tropical cyclones generated in the Bay of Bengal before the twentieth century that killed at least 100,000 people.

The Tropical Cyclone Bhola in 1970 that struck Bangladesh caused the largest number of deaths, specifically, up to 500,000 deaths (Ali 1999; Cerveny et al. 2017). The 1991 tropical cyclone Gorky that made landfall in Bangladesh was the third deadliest cyclone, killing around 140,000 people. The fourth largest number of fatalities also occurred in the BOB: the 2005 tropical cyclone that hit Myanmar and killed about 140,000 people.

The table also reveals that the Philippines in the Northwest Pacific Ocean, Region IV in the map, is one of the most fragile countries under the attacks of typhoons. Recently in 2013, Typhoon Haiyan killed 6,329 people in the country. Typhoon Haiyan, known in the country as Super Typhoon Yolanda, was widely reported at the time of its landfall to be the most intense cyclone in the recorded history, with 315 km/h maximum sustained wind speeds (Seo 2018). Typhoon Nina listed in Table 5.1 was the second deadliest tropical cyclone, generated in the Northwest Pacific Ocean, killing over 220,000 people in China and Taiwan in 1975.

5.4   What Causes Human Fatalities: Wind Speed or Storm Surge?

Although I introduced above that the intensity of a cyclone is measured by its wind speeds, specifically, the maximum sustained wind speeds (MSWS), as in the Saffir-Simpson Hurricane Scale, it can be measured by other measures too, e.g., minimum central pressure (MCP). An economic analysis showed that the MCP of a cyclone is a more accurate measure of the cyclone's intensity, considering its human and economic tolls (Bakkensen and Mendelsohn 2016). The lower the MCP at the eye of a cyclone, the more intense the cyclone becomes. Put differently, the larger the differential between the pressure at the cyclone's eye and the pressure at the boundary of the cyclone, the stronger the cyclone becomes.

An even more sophisticated scientific measure of hurricane intensity was also developed by scientists. A notable one is the Power Dissipation Index (PDI) which quantifies the intensity of a cyclone by tracing the entire history of the cyclone from an origination to an eventual disappearance, specifically, the full history of the maximum sustained wind speeds (Emanuel 2008).

Is the hurricane intensity measured by either maximum sustained wind speeds (MSWS) or minimum central pressure (MCP) what determines the degree of devastation caused by a hurricane? The devastation can be measured by the economic damage that is caused by the hurricane. It can be also measured by the number of human fatalities from the hurricane (Seo 2015). The cyclone scientists that have long studied the Indian Ocean suspected that there is another critical feature other than the MSWS and the MCP that causes a large number of human deaths: a storm surge (Murty et al. 1986; Dube et al. 2009).

The storm surge is the surge in the sea caused by the cyclone measured as the meters above the sea level. The higher the storm surge, the more dangerous the cyclone becomes to the areas where the cyclone makes landfall. This would be especially the case where the elevation of the landfall communities is low. Let's go back to Fig. 5.1 and confirm that the coastal areas of the Bay of Bengal are indeed very low-lying, with most areas located below 10 meters above the sea level.

A recent study attempted to quantify the relative importance of the storm surge in determining the number of human fatalities in the Indian Ocean, compared with the traditional cyclone intensity measures such as the maximum wind speeds or minimum central pressure (Seo and Bakkensen 2017). It finds that the storm surge is indeed the primary killer of people in the Indian Ocean.

Specifically, they find that a 10-cm increase in storm surge leads to a 14% increase in the number of fatalities. By contrast, a one-millibar (1 hPa) decrease in the minimum central pressure leads to a 7.3% increase in the number of fatalities. In a more sophisticated model that accounts for the dependence between the storm surge and the cyclone intensity, a 10-cm increase in the level of surge alone, after removing the effects of the minimum central pressure (MCP), leads to a 9.9% increase in the number of fatalities (Seo and Bakkensen 2017).

The research highlights the intersection of climate science and economics clearly. Although climate science has long focused on the cyclone power measures such as the maximum wind speeds and the minimum central pressure, as far as economic and human tolls are concerned, it is the height of storm surge that is more pivotal. In other words, unless you look at the problem from the viewpoint of economic damages as well as options to avoid the catastrophic events, it would remain unclear to you why the storm surge should be the most appropriate defining characteristic of tropical cyclones.

You might also be curious why the sea surge induced by a cyclone is more deadly than the cyclone power. In the low-lying coastal areas such as those in the river delta of Bangladesh, it is difficult for people to find a shelter against a sudden rise in the sea water caused by the cyclone. By contrast, people can more easily take a shelter against a high intensity storm by going down a cellar/bunker dug in preparation for intense cyclones. In a high storm surge, the surge of seawater would flood the cellar, making it more dangerous rather than safer. In the low-lying areas of the Ganges Delta, there are in many communities no high grounds, such as hilltops or rooftops of a multi-storied building, from which people can seek a refuge temporarily (Seo and Bakkensen 2017).

5.5   ADAPTATION: CYCLONE SHELTERS

After the catastrophic cyclone Bhola in 1970 that killed as many as 300,000 to 500,000 people in the lowlands of Bangladesh, concerned groups would have pondered over what would be the best strategy to avoid such a large-scale catastrophe again. Informal records on the tropical cyclone fatalities before the twentieth century indicate that deadly cyclones at the scale of cyclone Bhola may have struck half a dozen times the coastal communities in Bangladesh and India adjacent to the Bay of Bengal (Ali 1999).

If you were one of those concerned people, a simple solution might have come across your mind: Let's relocate the residents of the BOB lowlands to other areas on high grounds or further inlands. This simple and obvious solution had two major flaws. One is that, given that much of the Bangladesh is low-lying lands, it would not be easy to find alternative places for the mass relocation. Second, the Bengal Delta has the most fertile lands for agriculture upon which Bangladeshis rely for food and income. It would be difficult for the residents to desert the fertile lands for the less productive lands.

An alternative strategy was to construct a sufficiently large number of cyclone shelters in the coastal lowlands. The shelters had to be in a certain structure and meet geographical requirements. First, a shelter should be a three-story building in which the first floor is a piloti structure, the second floor is for animals and livestock, and the third floor is for fled people. Second, the shelters had to be built in a sufficiently large number to accommodate the large population in the densely populated coastal zones of Bangladesh. Third, the shelters should be built in a way to be accessible for nearly all people, which meant that the distances between one shelter to another should be kept within a certain limit.

As a matter of fact, the Bangladesh government initiated a cyclone shelter program (CSP) after the catastrophe of cyclone Bhola, which were followed by the supports from the international organizations such as the World Bank (Paul 2009). The CSP has built the cyclone shelters across the Bangladesh coast with each one accommodating 500 to 2,500 people. In particular, the World Bank started in 2007 to provide Bangladesh Government with the funds to build 230 new shelters and repair 240 existing shelters through the Emergency Cyclone Recovery and Restoration Project (World Bank 2008).

Has the construction of cyclone shelters turned out to be effective in reducing the number of human fatalities? A researcher compared the two cyclones that had similar physical and topological characteristics: Cyclone Gorky in 1991 and Cyclone Sidr in 2007 (Paul 2009). The two cyclones were both category 4 cyclones. Cyclone Gorky, as shown in Table 5.1, killed as many as 140,000 people. By contrast, Cyclone Sidr killed as many as 3,406 people. The researcher attributed the large reduction in the number of fatalities to the cyclone shelter program.

A more comprehensive study on the effectiveness of the cyclone shelter program was initiated recently, with a quantitative analysis. The author built the Negative Binomial (NB) model of the number of cyclone fatalities with the data of all the tropical cyclones in the Indian Ocean that made landfall during the 26-year period from 1990 to 2015 (Seo 2017). The NB model is simply the statistical model that explains the count data, in this case, the number of deaths in each tropical cyclone event, by a set of explanatory variables.

The study confirms that the cyclone shelter program succeeded in reducing the number of fatalities during the time period, assuming that the cyclone shelters were effectively and sufficiently established by the year 2007. Struck by the same height of a storm surge, the cyclone shelter program reduced the number of human deaths by 75%. In other words, the cyclone shelter program reduced the number of cyclone deaths from 100 to 25, given the same surge height.

The modeling proves that the cyclone shelters are highly effective against a high storm surge during a tropical cyclone landfall. It also tells us that, in other parts of the world with low-lying coastal areas that are cyclone prone, private housing developments should heed to the lesson of Bangladesh. Rather than building a single-story house, a multi-storied house is a far safer option.

From the public sector viewpoint, this in turn means that the building code and permits by the local government in the hurricane-afflicted zones are a key policy variable of adaptation against hurricane disasters (Knowles and Kunreuther 2014). In the low-lying zones frequently hit by hurricanes, a multi-story building, say, at least two, could be required as a precondition for new constructions.

From the market standpoint, a hurricane insurance market may be developed in a certain way that prices the premium at a higher level for the houses and buildings that are more likely to be devastated by hurricanes owing to a single-story structure, a closer location to the sea, and low altitude of the neighborhoods (FEMA 2012).

5.6   ADAPTATION OPTIONS: TECHNOLOGICALLY ADVANCED

Hurricanes, cyclones, and typhoons have wreaked havoc on coastal communities around the Planet from an unknown period of time. Scientists, however, began to record the full extent of the global cyclone occurrences only by the last quarter of the twentieth century and further became capable of modeling and predicting a cyclone generation and its characteristics only by the last decade of the century (McAdie et al. 2009).

Even during the first six decades of the twentieth century, scientists relied on coastal observations, island sightings, and ship reports to identify cyclone occurrences. As such, the cyclone data during these times, including frequency and intensity, were not precise. With the advance in the satellite recording of cyclones that became effective by the early 1970s, scientists were able to obtain precise global cyclone data (Landsea 2007).

The satellite recording of the hurricanes was an important milestone in the development of hurricane science but also in the humanity's capacity to deal with and adapt to future tropical cyclones. The satellite observations offer precise data of cyclones but also observe the genesis of a hurricane and its tracks in real time, equipping a governmental agency with the capacity to provide early information to coastal communities to be impacted.

Another important scientific advancement was the progress in the cyclone trajectory projection methods. A study of the cyclones generated in the Southern Hemisphere oceans revealed that the cyclone trajectory projection modeling began to be employed during the early 1990s in Australia. Since the cyclone path prediction models can inform coastal communities with an exact landfall location of a tropical cyclone several days before its eventual strike, it contributed largely to reducing the number of tropical cyclone fatalities dramatically in Australia (Seo 2015).

A review of the tropical cyclones generated in the Bay of Bengal since the 1990s shows that the trajectory projection modeling has evolved to an ever more advanced model. From the Limited Area Modeling (LAM) method, it evolved to the Quasi-Lagrangian Model (QLM) and then to the Non-hydrostatic Meso-scale Model (NMM) (Seo and Bakkensen 2017).

The cyclone trajectory prediction technology is, however, expected to have been more effective against high intensity cyclones, *i.e.*, the cyclones with high wind speeds. It might have been less effective against the cyclones with a high storm surge. For this reason and others, a cyclone storm surge modeling has become an important area of cyclone science (Hubbert et al. 1991; Dube et al. 2009). The prediction models of the impacts of climate change on tropical cyclones, say, in one or two centuries have concentrated on the impacts thereof on the frequency and intensity of cyclones, but it seems more urgent to assess the impacts of climate change on surge heights of future cyclones, especially in the Bay of Bengal neighbors (Knutson et al. 2010; Lin et al. 2012).

5.7   Chapter Highlights

- The world's oceans are a pivotal player in climate change. In particular, a warmer ocean may generate a tropical cyclone more frequently that is also more intense.
- The historical data shows that six of the eight deadliest cyclones globally occurred in the Bay of Bengal, with three of them killing over 100,000 people.
- The particularly high risk faced by the countries abutting the Bay of Bengal can be attributed to the fertile Bengal Delta for agriculture, high population density, and a topology of low-lying lands.
- The high fatality rate results from both cyclone intensity and storm surge, with the latter being a more dominant cause in the Bay of Bengal.

- This chapter highlights the efforts to minimize cyclone deaths in Bangladesh by constructing cyclone shelters carefully. The cyclone shelter program has reduced the number of deaths by 75% given the same height of storm surge.
- Adaptations to tropical cyclones under a warmer Planet would call for advanced technological options such as satellite monitoring, a cyclone trajectory projection, and a cyclone surge modeling.

References

Ali, A. 1999. Climate Change Impacts and Adaptation Assessment in Bangladesh. *Climate Research* 12: 109–16.

Bakkensen, Laura A., and Robert O. Mendelsohn. 2016. Risk and Adaptation: Evidence from Global Hurricane Damages and Fatalities. *Journal of the Association of Environmental and Resource Economists* 3 (3): 555–87.

Broecker, Wallace S. 1997. Thermohaline Circulation, the Achilles Heel of Our Climate System: Will Man-Made CO2 Upset the Current Balance? *Science* 278 (5343): 1582–88.

Cerveny, Randall S., Pierre Bessemoulin, Christopher C. Burt, Mary Ann Cooper, Zhang Cunjie, Ashraf Dewan, Jonathan Finch, et al. 2017. WMO Assessment of Weather and Climate Mortality Extremes: Lightning, Tropical Cyclones, Tornadoes, and Hail. *Weather, Climate, and Society* 9 (3): 487–97.

Church, J.A., P.U. Clark, A. Cazenave, J.M. Gregory, S. Jevrejeva, A. Levermann, M.A. Merrifield, G.A. Milne, R.S. Nerem, P.D. Nunn, A.J. Payne, W.T. Pfeffer, D. Stammer, and A.S. Unnikrishnan. 2013. Sea Level Change. In *Climate Change 2013: The Physical Science Basis*. Cambridge: Cambridge University Press.

Ciais, P., C. Sabine, G. Bala, L. Bopp, V. Brovkin, J. Canadell, A. Chhabra, R. DeFries, J. Galloway, M. Heimann, C. Jones, C. Le Quéré, R.B. Myneni, S. Piao, and P. Thornton. 2013. Carbon and Other Biogeochemical Cycles. In *Climate Change 2013: The Physical Science Basis*. Cambridge: Cambridge University Press.

Dube, S.K., Indu Jain, A.D. Rao, and T.S. Murty. 2009. Storm Surge Modelling for the Bay of Bengal and Arabian Sea. *Natural Hazards* 51 (1): 3–27.

Emanuel, Kerry. 2008. The Hurricane—Climate Connection. *Bulletin of the American Meteorological Society* 89 (5): ES10–ES20.

Federal Emergency Management Agency (FEMA). 2012. *Biggert-Waters Flood Insurance Reform Act of 2012 (BW12) Timeline*. Washington, DC: FEMA.

Feely, R.A., R. Wanninkhof, P. Landschützer, B.R. Carter, and JA. Triñanes. 2020. Global Ocean Carbon Cycle [in State of the Climate in 2019]. *Bulletin of the American Meteorological Society* 101 (8): S170–75.

Hubbert, Graeme D., Greg J. Holland, Lance M. Leslie, and Michael J. Manton. 1991. A Real-Time System for Forecasting Tropical Cyclone Storm Surges. *Weather and Forecasting* 6 (1): 86–97.

Johnson, G.C., J.M. Lyman, T. Boyer, L. Cheng, C.M. Domingues, J. Gilson, M. Ishii, R.E. Killick, D. Monselesan, S.G. Purkey, and S.E. Wijffels, 2020: Ocean Heat Content [in State of the Climate in 2019]. *Bulletin of the American Meteorological Society* 101 (8): S140–44. https://doi.org/10.1175/bams-d-20-0105.1.

Knowles, Scott Gabriel, and Howard C. Kunreuther. 2014. Troubled Waters: The National Flood Insurance Program in Historical Perspective. *Journal of Policy History* 26 (3): 327–53.

Knutson, T.R., J.L. McBride, J. Chan, K. Emanuel, G. Holland, C. Landsea, I. Held, J.P. Kossin, A.K. Srivastava, and M. Sugi. 2010. Tropical Cyclones and Climate Change. *Nature Geoscience* 3: 157–63.

Landsea, Christopher. 2007. Counting Atlantic Tropical Cyclones Back to 1900. *Eos, Transactions American Geophysical Union* 88 (18): 197–202.

Lin, Ning, Kerry Emanuel, Michael Oppenheimer, and Erik Vanmarcke. 2012. Physically Based Assessment of Hurricane Surge Threat Under Climate Change. *Nature Climate Change* 2 (6): 462–67.

McAdie, Colin J., Christopher W. Landsea, Charles J. Neuman, Joan E. David, Eric Blake, and Gregory R. Hamner. 2009. Tropical Cyclones of the North Atlantic Ocean, 1851–2006. Historical Climatology Series 6-2, Prepared by the National Climatic Data Center, Asheville, NC in cooperation with the National Hurricane Center, Miami, FL, 238p.

Murty, T.S., R.A. Flather, and R.F. Henry. 1986. The Storm Surge Problem in the Bay of Bengal. *Progress in Oceanography* 16 (4): 195–233.

National Hurricane Center (NHC). 2020. *Worldwide Tropical Cyclone Centers.* Miami, FL: NHC, National Oceanic and Atmospheric Administration.

Paul, Bimal Kanti. 2009. Why Relatively Fewer People Died? The Case of Bangladesh's Cyclone Sidr. *Natural Hazards* 50 (2): 289–304.

Revelle, Roger, and Hans E. Suess. 1957. Carbon Dioxide Exchange Between Atmosphere and Ocean and the Question of an Increase of Atmospheric CO_2 during the Past Decades. *Tellus* 9 (1): 18–27.

Saffir, Herbert S. 1977. *Design and Construction Requirements for Hurricane Resistant Construction.* New York: American Society of Civil Engineers.

Seo, S. Niggol. 2015. Fatalities of Neglect: Adapt to More Intense Hurricanes Under Global Warming? *International Journal of Climatology* 35 (12): 3505–14.

Seo, S. Niggol. 2017. Measuring Policy Benefits of the Cyclone Shelter Program in the North Indian Ocean: Protection from Intense Winds or High Storm Surges? *Climate Change Economics* 08 (04): 1750011. Accessed from https://doi.org/10.1142/s2010007817500117.

Seo, S. Niggol. 2018. Two Tales of Super-Typhoons and Super-Wealth in Northwest Pacific: Will Global-Warming-Fueled Cyclones Ravage East and Southeast Asia? *Journal of Extreme Events* 05 (02n03): 1850012. Accessed from https://doi.org/10.1142/s2345737618500124.

Seo, S. Niggol, and Laura A. Bakkensen. 2017. Is Tropical Cyclone Surge, Not Intensity, What Kills So Many People in South Asia? *Weather, Climate, and Society* 9 (2): 171–81.

United Nations. 2017. *Fact Sheet. The Ocean Conference.* New York, NY: UN.

World Bank. 2008. *Bangladesh—Emergency 2007 Cyclone Recovery and Restoration Project.* Washington, DC: The World Bank. Accessed from http://documents.worldbank.org/curated/en/763581468013246012/Bangladesh-Emergency-2007-Cyclone-Recovery-and-Restoration-Project.

CHAPTER 6

Sublime Grasslands: A Story of the Pampas, Prairie, Steppe, and Savannas Where Animals Graze

6.1   MAJESTIC GRASSLANDS

Planet Earth is home to many iconic grasslands. You may have come, via varied roads, to intimate familiarity with the Prairie in North America, or the Pampas in South America, or the Llanos in the Andes, or the Steppe in Central Asia, or the Savannas in Africa, or the rangelands in Australasia. You may have in fact grown up in a little house in the Prairie or in a Mongolian tent in the Eurasian Steppe (Wilder 1932).

A grassland is one of the many biomes or ecosystems in the world. A biome is a community of plants and animals with many commonly shared characteristics. An ecosystem is a similar concept, a derived term from the concept of a biome after putting an emphasis on climate and soil characteristics (Matthews 1983; Box and Fujiwara 2005).

The magnificent grasslands of the Planet have received a great deal of attention from climate researchers. The Sahel and West Africa in Sub-Sahara, for example, is one of the most researched areas in the world by climate researchers. Like the Sahel, major grasslands are bordered by equally majestic deserts of the world, say, the Sahara Desert. Unlike the deserts, the Sahelian and Savanna regions are inhabited by a community of humans who manages natural resources endowed for the grasslands.

The geography of the grasslands lies between the desert geography and the cropland geography in the climate sphere. The grasslands receive not sufficient rainfall for growing grains and vegetables, as such, characterized by arid/semi-arid zones. Nevertheless, the grassland regions receive small but sufficient rainfall to support a community of grasslands. The grassland biome/ecosystem is defined by the land areas dominated by grasses of various types: short, medium, and tall grasses.

To the researchers who were concerned about food security caused by global warming, the grasslands like the Sahel and Savannas of Sub-Sahara were treated, though unintentionally, as barren lands where crops and cereals will fail routinely (Seo 2014a, b). Gradually, it has dawned on the climate researchers that the grasslands are vital places of living, both economically and ecologically.

Ecologically, a grassland is home to a community of different types of grasses that offer many much valuable ecological functions. A climate benefit is one of the valuable ecological functions of the grasslands. Of the well-studied climate benefits of the grasslands is a carbon sequestration from the atmosphere. Another is a preservation of soil carbons.

Economically and most prominently, the grasslands support the world's vast farm animals and livestock productions (Seo 2006). The Pampas in South America, for example, support the world's largest production of cattle across Argentina, Brazil, and Uruguay. The Great Plains in the US support the productions of different species of livestock. The rangelands in Australia and New Zealand support a large industry of cattle and other animals oriented to exports to Asia.

Other than beef cattle, most frequently raised animals across the Planet are, inter alia, dairy cattle, buffaloes, water buffalos, sheep, goats, pigs, chickens, turkeys, beehives, donkeys, dogs, horses, camels, American bison, Himalayan yak.

These animals are the biggest supplier of nutrition to the humanity through their meat, eggs, and milk (OECD 2019). They provide vital energy (calorie) and essential micro-nutrients (say, vitamins, calcium, iron, etc.) for the humanity's survival. In addition, they offer wools of various types which are turned into popular shopping items, e.g., ugg boots, cashmere sweaters.

Of the farm animals, some are a ruminant: cattle, goats, sheep, buffaloes, yak, camels. The ruminant is an animal whose digestive function relies on four rumens for the rumination process. The digestive process results in the production and eventually emissions of methane (CH_4) gas into the atmosphere, through mostly burping and farting. The methane is one of the most potent greenhouse gases (FAO 2013; GCP 2016; Schaefer et al. 2016).

An emerging frontier of climate science is to developing techniques to reduce or capture methane emissions from ruminant animals. This effort is sometimes dubbed the "future feed" project. As it indicates, the research frontier is formed around finding an alternative feed for ruminant animals, the digestion of which do not result in or help reduce the production of methane gas (Hristov et al. 2015). Another way to proceed in this effort is to capture methane emissions after the rumination process, say, a post-rumination technology. For example, a methane capture may be installed in the special corrals that keep ruminant animals or through a burp-catching mask (Wired 2021).

6.2   THE PLANET'S ICONIC GRASSLANDS

On Earth, a grassland is the dominant land biome, covering 40.5% of the land surface of the Planet (WRANGLE 2020). Figure 6.1 shows where the Planet's prime grasslands are located. From the global land cover data provided by the National Oceanic and Atmospheric Administration (NOAA), the figure highlights regions of forbs, short grasses, medium

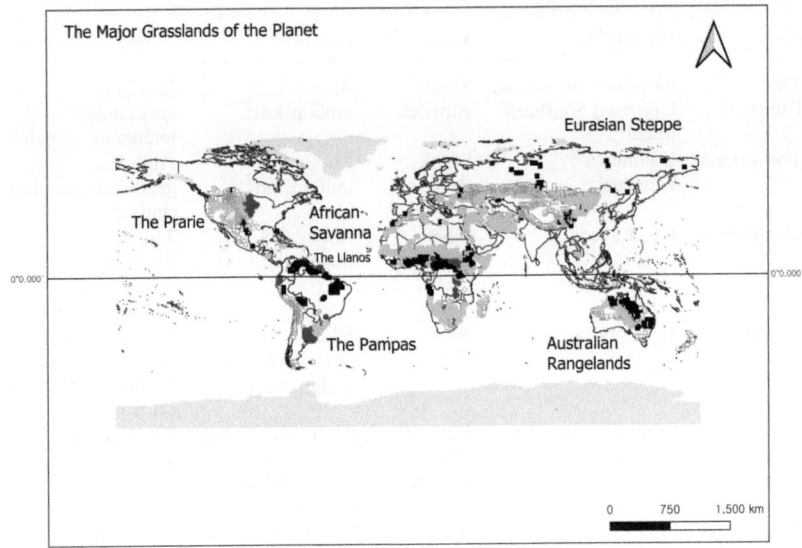

Fig. 6.1   The major grasslands of the Planet

grasses, and tall grasses (Matthews 1983). The six iconic grasslands are marked: The Pampas, the Llanos, the Steppe, the Prairie, the African Savannas, and the rangelands of Australasia. Let us begin our analysis with taking a rather detailed look at each of these great grasslands.

6.2.1   The Pampas

The first grabbing our attention is the Pampas in South America. The Pampas, which means the plains in Spanish, are lowland grasslands located in five northern provinces of Argentina, all of Uruguay, and the southern-most state of Brazil. The lowland plains stretch vastly across southern South America over the area of 1.2 million square kilometers (see Table 6.1), rarely interrupted by high hills or mountains (Seo 2012a). For your reference, the entire state of Texas is as large as about 700,000 km^2.

As far as the climate is concerned, the Pampas belongs to a temperate climate zone. The rainfall across the lowland plains is varied: 50 milli meters per month in the semi-arid regions and 100 milli meters per month

Table 6.1   Profiles of the major grasslands

ID	Geography	Continent	Size (km^2)	Biomes and climate
The Pampas	Northern Argentina, Uruguay, Southern Brazil	South America	About 1.2 million km^2	Lowland grasslands; temperate climate
The Llanos	Colombia, Venezuela	South America	About 0.57 million km^2	Highland grasslands; tropical climate
The Prairie	US, Canada	North America	About 2.8 million km^2	Tallgrass, shortgrass, mixed-grass; temperate climate
African Savannas	Sahel and across Africa	Africa	50% of Africa is rangelands (15 million km^2)	Dry savanna, moist savanna; tropical/subtropical climate
Eurasian Steppe	From Hungary to Mongolia	Eurasia (Asia, Europe)	About 8,000 km long; about 4.5 ~ 5.5 million km^2 (est.)	Short grasses; temperate arid climate
Rangelands	Australia (outback), New Zealand (Tussock grasslands)	Australasia	81% of Australia (6.2 million km^2) is rangelands	Grasslands, savannas, shrubs; arid and semi-arid climate

in the moist zones. The soils are fertile especially for the growth of grasses (WRANGLE 2020).

6.2.2   The Llanos

The Llanos is another South America's iconic grassland biome, which is salient as a highland grassland. The Llanos is delimited by the Andes Mountains to the north and west and by Guiana Highlands to the east. The elevations thereof rise from Llanos Bajos (Low Plains in English) to Llanos Altos (High Plains in English). Geopolitically, the Llanos stretches across northeastern Colombia and western Venezuela, whose size is roughly 570,000 km^2.

Because it is located near the Equator, the Llanos is a tropical grassland. With a high and concentrated rainfall pattern typical of the tropical climate regime, it is also a flooded grassland. The monthly rainfall ranges from 90 mm in the low rainfall zones to 380 mm in the high rainfall zones (WWF 2020).

6.2.3   The Prairie

The Prairie refers to the varied grasslands in central North America, whose geographical boundaries contain the Great Plains. The Prairie borders the forested lands to the east and the arid desert-like lands to the west. The amount of rainfall across the Prairie varies from 90 mm/month in the eastern edge to 25 mm/month in the western edge.

The vegetations of the Prairie are dominantly perennial grasses and flowering plants. Three sub-regions of the Prairie are identified by the type of the grasses, as marked in Fig. 6.1: tallgrass Prairie, midgrass or mixed-grass Prairie, shortgrass Prairie. The grass types of the three sub-regions are determined by, *inter alia*, the amount of rainfall (WRANGLE 2020).

Before the arrival of European settlers, the Prairie had been roamed by wild animals, most notably, bison and wolves. The bison is also known as a buffalo in the US and Canada, but is only remotely related to a true buffalo, e.g., the water buffalo in India that I explained in Chapter 4. The magnificent landscape of the bison herd roaming the Prairie had long disappeared, but some conservationists are working on restoring them by purchasing private lands (Rosner 2012).

6.2.4   African Savannas

The savanna refers to the landscapes or vegetation types that have a continuous grass cover, interrupted occasionally by trees and shrubs. The largest area of the savanna biome is found in Africa, of which the Sahel and the Sudanian Savanna below the Sahara Desert are famed most. Beyond the two, about a half of the African continent is grasslands including the Veld grasslands in the south of the continent (WRANGLE 2020).

The climate of the African savanna zones is characterized by a hot tropical/subtropical climate as well as a long dry season. Depending upon the length of the dry season, the savanna zones can be divided into different types, for example, moist savanna and dry savanna. The Sahel belongs mostly to the dry savanna while the Sudanian Savanna belongs to the moist savanna.

In the dry savanna zones in the Sahel, the dry season may last as long as 11 months during the year. The annual rainfall can fall as low as 500 milli meters, that is, about 40 milli meters per month, which makes a farming of staple grains, such as rice and wheat, extremely difficult (Seo 2014b).

6.2.5   The Steppe

The Steppe refers to the grassland regions in Asia and Europe. The Steppe grassland belt extends 8,000 km from the east to the west: Hungary from the west, Ukraine and Central Asia in the middle, and Mongolia and Manchuria in the east. Mountain ranges interrupt the Steppe grasslands.

The Steppe, meaning a flat grassy plain in Russian, is an ecological zone characterized by grassland plains without trees. The grasses are short grasses, as marked in Fig. 6.1, owing to an arid climate while trees and shrubs in the Steppe are found only near lakes and rivers. The Steppe ecosystem is similar to the shortgrass Prairie in North America (Fig. 6.1) (NGS 2020).

Throughout history, Steppe peoples interacted through the entire range of the grasslands because horsemen could easily ride across the shortgrass plains, crossing national boundaries, with a portable round tent called Yurt. Notably, The Mongol empire founded by Genghis Khan was able to conquer the entire range of the Steppe as well as the Chinese empire at the time, via their superior cavalry.

6.2.6   Rangelands in Australia and New Zealand

Australia and New Zealand, the two largest countries in the South Pacific Ocean, are home to major grasslands and simultaneously famed for their livestock productions. The grasslands thereof are better known as the outback in Australia and tussock grasslands in New Zealand. First European settlers in the two countries earned income through sheep farming, selling meat, wool, and skins, which still holds cultural importance through, for example, national shearing championships and ugg boots.

The outback of Australia is also referred to as rangelands. The rangelands cover about 81% of the country which is a vast country (Table 6.1). The rangelands are arid and semi-arid regions of the country, whose biomes are dominantly grasslands, savannas, and shrublands. Across the rangelands, the amount of rainfall is scarce, below 30 mm/month, about 350 mm/year (ACRIS 2008; Seo 2011).

For New Zealand, pastoral farming is the dominant economic activity in the rural areas. It accounts for about 60% of employment while the rest is accounted for by crop farming, forestry, horticulture, and aquaculture. The major animals for pastoral farming are beef cattle, dairy cattle, and sheep. The pastoral farming of the country is predominantly based on grazing grasses and pastures, rather than hay or silage.

The native grasslands of New Zealand are known as tussock grasslands (Mark 2007). Tussocks are grasses that grow in the form of a clump. Tussocks are of two types: tall tussocks and short tussocks. Farm animals can feed on short tussocks, but not tall tussocks. When Europeans settled, they burnt native tussocks and planted exotic grasses for grazing farm animals.

Pastures were created to grow grasses and plants for grazing animals. Many different types of grasses and other plants such as legumes and clovers are grown for the pastures. Pastoral farming in New Zealand rely on pastures, not tussock grasses that were relied upon by the early European settlers.

6.3   Climate Benefits: Carbon Sink and Preservation of Soil Carbon

The majestic grasslands of the Planet which extend vastly offer numerous climate and ecological benefits that have attracted attention. Of the climate benefits, two are noticeable as a climate change mitigation option. The first is their role as a sink of carbon dioxide through the photosynthesis process and the second is their role of preventing soil carbons from being released.

Let's begin with a basic biochemistry of plants. Most plant species, about 90%, are C-3 plants which rely on C-3 photosynthesis, that is, 3-carbon, including trees, shrubs, and crops. C-4 plants include most tropical/subtropical grasses and several major crops such as maize, millet, sorghum, sugarcane, and switchgrass. A legume (soybean) is another category of plants, called CAM plants which include succulents (Reich et al. 2018).

An increase in carbon dioxide in the atmosphere is beneficial to the growth of grasses. A meta-analysis of 15 years' FACE experiments shows that a doubling of carbon dioxide concentration in the air results in an increase of leaf area index by about 10% as well as an increase of dry matter production by about 10%, both for C-3 grasses. For C-4 grasses, the dry matter production has increased by 40% in response to the same change. For legumes, the dry matter production has increased by 20% in response to the same change in the CO_2 concentration (Ainsworth and Long 2005).

The grasses (and plants in general) grow larger because an increase in carbon dioxide enhances their photosynthesis process through which carbon dioxide is absorbed from the air. How much carbon dioxide do the grasses absorb? Is it a significant amount to be considered as a climate change mitigation option? The perennial grasslands are estimated, from the Australian perennial grasslands, to sink 5 tons of carbon at the "present" climate, equivalently about 20 tons of carbon dioxide, annually per hectare of the grasslands (Christie 1981).

If the atmospheric carbon concentration should double, how will the carbon uptake of grasses be affected? The aforementioned meta-analysis shows that the diurnal carbon uptake increases by nearly 40% for grasses from the present level. For legumes, the diurnal uptake of carbon increases by about 20% from the present level in response to the doubling of carbon dioxide (Ainsworth and Long 2005).

These predictions are averages of a large number of predictions, each of which is obtained from an individual field experiment. An individual experiment's prediction on an indicator can be substantially different from the average prediction, depending on, inter alia, how accurately the field experiment is designed and set up (Shaw et al. 2002). An individual experiment's prediction can be far apart from the average prediction owing to its length of the observation period, *e.g.*, 20 years instead of 10 years (Reich et al. 2018). In particular, the 20-year FACE experiment argued that the yield increase of C-4 plants, which include tropical/subtropical grasses, exceeds the yield increase of C-3 plants after 12 years of observation period.

The second important role that grasslands play in climate mitigation is the preservation of soil carbons. The soils on the Planet store a massive amount of carbon, a disturbance of which releases carbon into the atmosphere (Refer to Fig. 3.2 in Chapter 3). To give you a visual example, a conversion of an old grassland into a crop land for food production will release soil carbons while a conservation of the old grassland against the conversion into a crop land would preserve soil carbons kept under the grasses.

From the Planet level, scientists estimate that the soils of the entire Planet store 1,500–2,400 PgC (Petagrams of Carbon), which is 1.5–2.4 trillion tons of carbon or alternatively 1,500–2,400 Giga tons of carbon (Ciais et al. 2013). Vegetations on the Earth surface, especially dense grasses, keep soil carbons from being released. You would wonder: Can we estimate the soil carbon storage at the grassland plot level? How much carbon is stored in one hectare of dense grassland? According to one study in Australia, one hectare of perennial grasslands in the high rainfall areas of Australia is estimated to preserve 72.9 tons of carbon (Chan and McCoy 2010).

How substantial is the soil carbon storage of grasslands? The size of rangelands in Australia is 623 million hectares, as shown in Table 6.1 in square kilometers, a fraction of which is perennial grasslands. The Pampas is as large as 120 million hectares while the Llanos is as large as 57 million hectares. The North American Prairie is of the size of 280 million hectares. The Eurasian Steppe is at least 260 million hectares and perhaps as large as 500 million hectares. About half of African continent is rangelands which includes arid zones, dry savannah, and moist savannah, which corresponds to about 1,500 million hectares (WRANGLE 2020).

Considering the vast extension of grasslands across the Planet, the soil carbon storage capacity is certainly very large and no one may doubt its significance in the global efforts to deal with carbon dioxide problem. This would be true even after considering that not the entirety of the aforementioned grasslands supports a perennial grassland. Consider also that shortgrasses are likely as effective as tallgrasses in preserving soil carbons.

6.4   LIVESTOCK MANAGEMENT

Across the Planet's premier grasslands, the dominant economic activity is livestock management. The Pampas grasslands in South America support the world's largest production of beef cattle in Argentina, Brazil, and Uruguay. The rangelands in Australia and New Zealand make it possible for the countries' massive exports of cattle and sheep for profits to Asian countries. The Eurasian Steppe has long been famed for livestock grazing including horses and cattle. The Great Plains in the North American Prairie is the backbone of the livestock industry of the US. The African Savannas are home to many different species of animals, but also support the livestock management of the Africans through which income as well as foods are earned.

Why is the livestock management favored on grasslands? There are many compelling reasons (Seo and Mendelsohn 2008; Seo 2014b). First, farm animals can graze grasses that grow abundantly on grasslands. A pasture is the grasslands for the grazing of farm animals. On the pasture, farm animals graze many different species of grasses and legumes which are called forages. That is, of the forages, grass forages and legume forages are by far most abundant (Murphy 1998). A forage crop can also be grown and sold for the markets, e.g., Californian alfalfa. Fodders are, on the other hand, an animal food given to the animals by humans such as hay, silage, etc.

Second, the grassland biome is characterized by arid and semi-arid climate zones in which the amount of precipitation is too little to grow major staple grains such as rice, wheat, and maize. The staple grains require abundant rainfall during the growing season, in addition to ample sunlight. Well aware of this constraint, farmers have little reason to choose to manage these grains on a large scale on the grasslands.

Third, an arid climate regime that prevails in the grassland biome makes livestock farms less vulnerable to numerous livestock diseases which often kill farm animals en mass. Livestock diseases can break out more easily and become highly contagious in a humid climate condition. I already discussed in Chapter 3 the problem of livestock diseases in Sub-Saharan Africa, in particular, the trypanosomiasis commonly known as sleeping sickness carried by tsetse flies (Ford and Katondo 1977).

What values does the livestock management offer to the farmers and the human society? If you happen to dislike the odors of livestock manure in rural areas, you may ask whether we can do away with raising farm livestock. If, on the other hand, you happen to enjoy a pastoral scenery in rural areas, it might not be difficult for you to imagine the values of livestock farming.

First of all, a livestock farming provides sources of income to farmers in rural areas. Even in the US, livestock sales, animal products included, account for about half of the farm income. The livestock income becomes a critical means of living through generations of both income and foods in a harsh climate zone such as the Sahel and the monsoon-afflicted Indian regions (Seo 2014b).

More critically, farm animals provide the humanity essential nutrients and energy that are needed to sustain a human body and bodily functions. The energy needed for bodily functions, measured in calorie, is obtained mostly by consuming meats, milk, and eggs from farm animals. From the consumptions of livestock products, humans obtain protein, lipids, carbohydrates, vitamins, calcium, iron, and other nutrients.

A recently added value is from climate change. A series of studies indicates that livestock management will be at the heart of a resilient agricultural portfolio in an era of climate change, which was elucidated in Chapter 2 of this book. A hot and dry climate regime will make farmers to choose livestock management more often against a specialized crop portfolio (Seo 2015). A more variable rainfall regime will make farmers to choose more frequently a mixed crop-livestock portfolio instead of a specialized crop or a specialized livestock portfolio (Seo 2012b). An increase in diurnal temperature range in farming regions will make farmers opt for farm animals, especially cattle, against the range of crops which would fail more easily in an increased daily temperature variation (Seo 2012b).

Besides the climate change push, there is another great force that pushes farmers in developing countries to switch to livestock portfolios. The meat and milk consumption per capita of underdeveloped countries, to be explained below presently, will increase as their income levels will get ever higher, as have been witnessed in developed countries as well as fast-growing economies. The population size thereof is expected to increase. This will provide an additional incentive for the livestock switch.

The relationship between the meat consumption per capita and the income per capita can be seen from Fig. 6.2. The meat consumption per capita on the horizontal axis is the sum of individual meat consumptions of beef, pork, poultry, and sheep (OECD 2019). In the richest country,

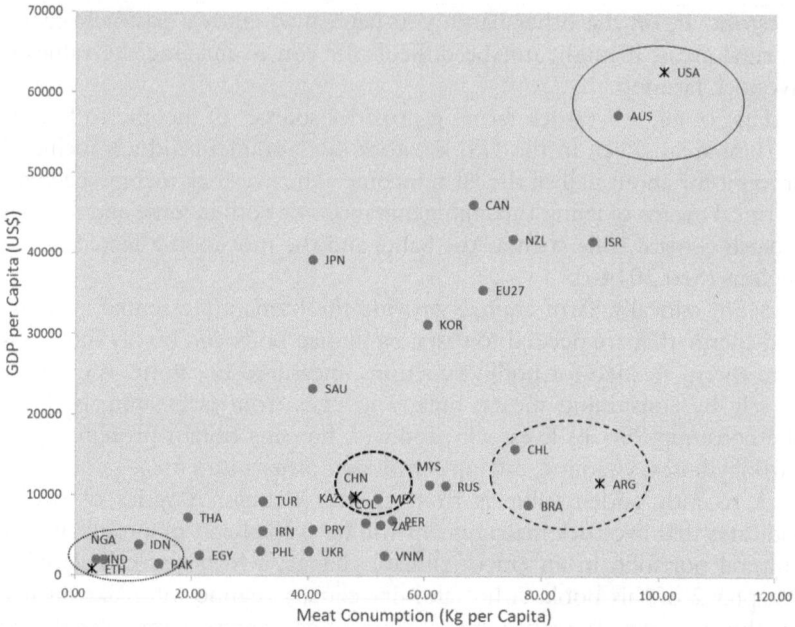

Fig. 6.2 Meat consumption per capita versus GDP per capita (2019)

the US with the GDP per capita of over US$63,000, the meat consumption is as large as 100 kg per capita annually. In the low-income countries such as Ethiopia, Nigeria, India, Indonesia, and Pakistan in the figure, the meat consumption is more or less 5 kg per capita annually. In China where the GDP per capita is about 10,000 US$, the meat consumption per capita is as large as about 50 kg per capita.

Note also that there is another cohort of countries in which the individual meat consumption is very large, despite the low level of personal income: Argentina, Brazil, and Chile (Seo 2015). For Argentina, the personal income is about 10,000 US$, but the personal meat consumption reaches 90 kg per year. This is most likely attributed to the vast grassland resources of the Pampas.

6.5   Methane Reduction Micro-Technologies

The meat consumption trend similar to Fig. 6.1 would be regarded as good news to underdeveloped countries: as their economies improve, their consumptions of meat which provide essential energy and numerous micro-nutrients will increase. To the climate activists, the trend is accepted as dreadful news. The source of their fear is the fact that the methane molecule (CH_4) which is emitted from livestock is a very powerful Earth-heating gas. It is 25 times more potent than carbon dioxide in its power to warm the Planet (GCP 2016). They are concerned that an increased meat consumption by the world's poor people will result in an increase in methane in the atmosphere (Schaefer et al. 2016).

If every country on the Planet were to consume meat like the US, as Fig. 6.2 indicates, the world meat consumption should increase by a lot more than 10 times, that is, 1,000%, the current level which would unavoidably lead to a massive increase in methane buildup owing to a large expansion of livestock productions across the vast grasslands surveyed in this chapter. For reference, livestock methane emissions at present amounts to as much as 7 GtCO2 which is about 15% of all anthropogenic greenhouse gas emissions (GCP 2016).

As elaborated in Sect. 6.3, livestock management can offer a mitigation opportunity of global warming by way of an increase in carbon dioxide uptake by grasses as well as the grasslands' preservation of soil carbons (Seo 2020). So, livestock management is not a blanketly bad news in regard to mitigation endeavors of the Planet.

Further, there is a silver lining, mostly unnoticed by concerned people, in the fact that the livestock methane emissions is a major contributor to the global increase in the emissions of greenhouse gases. Concretely, a reduction in animal methane emissions may be far easier technologically than other advanced technological options such as a direct carbon air capture or solar radiation reflector (Seo 2013). I once called the former as a micro-technology in contrast to the latter a global technology. The list of the advanced technological options will be explained in Chapter 9 of this book.

What technologies are available for reducing the methane emissions from farm animals, more specifically, ruminant animals? Three options are being mentioned most often and studied by scientists. One is to find alternative feeds. The second is the development of microbes that can eat methane. The third is to the switch to non-ruminant animals which do not emit methane.

The alternative feed approach is to find an alternative feed to farm animals the consumption of which results in much reduced or near zero emissions of methane from ruminant animals. The methane emissions from ruminant animals are released mostly from the rumination process, that is, burping, belching, and farting. Recently, researchers suggested seaweeds as an alternative livestock feed (Hristov et al. 2015).

The second approach is to develop microbes that can "eat" methane. Observed in the warmer Arctic permafrost which is melting due to the Planet warming, methane-oxidizing bacteria abundant in the organic rich Arctic soils are reported to reduce the Arctic methane emissions by half (Oh et al. 2020). Similarly, microbes may be developed in the context of ruminant animals.

The third is a non-ruminant animal approach. The low or near zero methane emissions by non-ruminant animals has received attention for a long time by some scientists, but largely been put aside up until now in the climate policy discussions (Kempton et al. 1976). Among the non-ruminant animals which are consumed by humans somewhere everyday are, inter alia, kangaroos, chickens, turkeys, and fish.

6.6   Becoming a Vegetarian for the Sake of Earth

Having come as far as we have thus far in this chapter, I feel we are pretty well positioned to ask a fundamental question regarding the livestock management often raised by young enthusiasts: Should we change our long-held food consumption behaviors to become a vegetarian or a vegan for the sake of the Planet? You notice that, in the above section, the third option that I proposed was to switch our consumption of meat from ruminant animals to non-ruminant animals. Then, you may ask why not switch all together to a vegetarian, that is, a non-meat-eater?

To refresh your memory, a vegetarian is someone who does not eat any animal meat including fish. A vegan is even more restrictive in consumption behaviors. In addition to animal meats, a vegan does not consume any animal products such as dairy milk, dairy cheese, and eggs.

Vegetarianism has long been a way of life for certain peoples. It is often culturally handed down, a glimpse of which can be seen also in Fig. 6.2. Notice that the individual meat consumption of India is among the lowest in the world. The Indian culture regards water buffaloes as their mother in their previous lives that gives milk to them in the present life, under which belief Indians do not consume the meat of water buffaloes and instead bury them when they die. In addition, many spiritual traditions in India, including Buddhism and Jainism, adhere to vegetarianism in the monastic communities.

The fraction of vegetarians/vegans among the population in India is as high as 31%, the highest of any nations in the world (Census of India 2014). For reference, a Gallup poll in 2018 reported that as high as 5–8% of the US population consider themselves as either a vegetarian or a vegan (Gallup 2018).

Considering the deep root of vegetarianism in long-held traditions, I wonder how many people have actually changed their diet from meat-eaters to non-meat-eaters because of the concern on global warming and the Planet. There is no data on this, as far as the present author is aware of. However, you will find easily a vegetarian friend around you who argues passionately as one of the benefits of adopting a vegetarian diet a contribution to reduced global warming.

The argument will become weaker if the adoption of a vegetarian/vegan diet should lead to increased consumptions of other types of food such as grains, vegetables, tubers, and root crops, all of which are releasing carbon and other greenhouse gases during their productions. Further, for the supply of adequate energy and nutrients that a human body needs, a vegetarian/vegan may also have to rely on nutritional supplements, the productions of which result in carbon emissions.

My brief review in this section cannot place an exact value of vegetarian/vegan diets in terms of global warming mitigation. At the moment of this writing, I would conclude that it should be left as an individual's lifestyle decision rooted on culture, belief, and perhaps health benefits.

6.7 CHAPTER HIGHLIGHTS

- The Planet is home to many famed grasslands, including the Pampas, the Llanos, the Prairie, the Steppe, the Savannas, and the Australian rangelands.
- The biome of grasslands, which are of various types, covers 40% of the Planet's lands.
- The two connections between climate change and the grasslands are: carbon dioxide sinks by grasslands and preservation of soil carbons.

- The grasslands offer valuable resources for livestock management: pastures and forages.
- The global livestock production is expected to increase in a major way as underdeveloped economies would grow in the coming decades and consume more meats, which has made climate activists concerned because of methane emissions from farm animals.
- This chapter explains three novel methane reduction technologies: alternative feeds, methane-eating microbes, switches to non-ruminant animals.
- A vegetarian diet is most often rooted in culture and belief, whose contribution to global warming mitigation is hard to measure.

REFERENCES

Ainsworth, E.A., and S.P. Long. 2005. What Have We Learned from 15 Years of Free-Air CO2 Enrichment (FACE)? A Meta-Analytic Review of the Responses of Photosynthesis, Canopy Properties and Plant Production to Rising CO2. *New Phytology* 165: 351–71.

Australian Collaborative Rangeland Information System (ACRIS). 2008. *Rangelands 2008—Taking the Pulse*. Canberra, AU: Department of Environment.

Box, E.O., and K. Fujiwara. 2005. Vegetation Types and Their Broad-scale Distribution. In *Vegetation Ecology*, ed. E. van der Maarel, 106–28. Oxford, UK: Blackwell Scientific.

Census of India. 2014. *Sample Registration System Baseline Survey 2014*. New Delhi, IN: Government of India.

Chan, K.Y., and D. McCoy. 2010. Soil Carbon Storage Potential Under Perennial Pastures in the Mid-north Coast of New South Wales, Australia. *Tropical Grasslands* 44: 184–91.

Christie, E.K. 1981. Biomass and Nutrient Dynamics in a C 4 Semi-Arid Australian Grassland Community. *The Journal of Applied Ecology* 18 (3): 907–18. https://doi.org/10.2307/2402381.

Ciais, P., C. Sabine, G. Bala, L. Bopp, V. Brovkin, J. Canadell, A. Chhabra, R. DeFries, J. Galloway, M. Heimann, C. Jones, C. Le Quéré, R.B. Myneni, S. Piao, and P. Thornton. 2013. Carbon and Other Biogeochemical Cycles. In *Climate Change 2013: The Physical Science Basis*. Cambridge: Cambridge University Press.

Food and Agriculture Organization (FAO). 2013. *Tackling Climate Change through Livestock: A Global Assessment of Emissions and Mitigation Opportunities*. Rome: FAO.

Ford, J., and K.M. Katondo. 1977. Maps of Tsetse Fly (Glossina) Distribution in Africa, 1973, According to Subgeneric Groups on a Scale of 1: 5000000. *Bulletin of Animal Health and Production in Africa* 15: 187–93.

Gallup. 2018. *Snapshot: Few Americans Vegetarian or Vegan.* Washington, DC: Gallup. Accessed from https://news.gallup.com/poll/238328/snapshot-few-americans-vegetarian-vegan.aspx.

Global Carbon Project (GCP). Global Methane Budget 2016. GCP. Accessed form https://www.globalcarbonproject.org/methanebudget/index.htm.

Hristov, Alexander N., Oh Joonpyo, Fabio Giallongo, Tyler W. Frederick, Michael T. Harper, Holley L. Weeks, Antonio F. Branco, et al. 2015. An Inhibitor Persistently Decreased Enteric Methane Emission from Dairy Cows with No Negative Effect on Milk Production. *Proceedings of the National Academy of Sciences, USA* 112 (34): 10663–68.

Kempton, T.J., R.M. Murray, and R.A. Leng. 1976. Methane Production and Digestibility Measurements in the Grey Kangaroo and Sheep. *Australian Journal of Biological Sciences* 29 (3): 209–14.

Mark, A.F. 2007. Grasslands. Te Ara—The Encyclopedia of New Zealand. Accessed on http://www.TeAra.govt.nz/en/grasslands from May 16, 2020.

Matthews, Elaine. 1983. Global Vegetation and Land Use: New High-Resolution Data Bases for Climate Studies. *Journal of Climate and Applied Meteorology* 22 (3): 474–87.

Murphy, B. 1998. *Greener Pastures on Your Side of the Fence: Better Farming with Voisin Management Intensive Grazing.* Colchester: Ariba Publishing.

National Geographic Society (NGS). 2020. *Steppe.* Washington, DC: NGS.

Oh, Y., Qianlai Zhuang, Licheng Liu, Lisa R. Welp, Maggie C.Y. Lau, Tullis C. Onstott, David Medvigy, Lori Bruhwiler, Edward J. Dlugokencky, Gustaf Hugelius, Ludovica D'Imperio, and Bo Elberling. 2020. Reduced Net Methane Emissions due to Microbial Methane Oxidation in a Warmer Arctic. *Nature Climate Change* 10: 317–321.

Organisation for Economic Co-operation and Development (OECD). 2019. *OECD-FAO Agricultural Outlook 2019–2028.* Paris, FR: OECD.

Reich, Peter B., Sarah E. Hobbie, Tali D. Lee, and Melissa A. Pastore. 2018. Unexpected Reversal of C3 Versus C4 Grass Response to Elevated CO2 During a 20-Year Field Experiment. *Science* 360 (6386): 317–20.

Rosner, H. 2012. *Dreaming of a Place Where the Buffalo Roam.* New Haven, CT: Yale Environment 360.

Schaefer, H., S.E.M. Fletcher, C. Veidt, et al. 2016. A 21st Century Shift from Fossil-fuel to Biogenic Methane Emissions Indicated by $^{13}CH_4$. *Science* 352: 80–84.

Seo, S. Niggol. 2006. Modeling Farmer Responses to Climate Change: Climate Change Impacts and Adaptations in Livestock Management in Africa. PhD dissertation, Yale University, New Haven, CT.

Seo, S. Niggol. 2011. The Impacts of Climate Change on Australia and New Zealand: A Gross Cell Product Analysis by Land Cover. *Australian Journal of Agricultural and Resource Economics* 55 (2): 220–38.

Seo, S. Niggol. 2012a. Adaptation Behaviours Across Ecosystems Under Global Warming: A Spatial Micro-Econometric Model of the Rural Economy in South America. *Papers in Regional Science* 91 (4): 849–71.

Seo, S. Niggol. 2012b. Decision Making Under Climate Risks: An Analysis of Sub-Saharan Farmers' Adaptation Behaviors. *Weather, Climate, and Society* 4 (4): 285–99.

Seo, S. Niggol. 2013. Economics of Global Warming as a Global Public Good: Private Incentives and Smart Adaptations. *Regional Science Policy & Practice* 5 (1): 83–95.

Seo, S. Niggol. 2014a. Adapting Sensibly When Global Warming Turns the Fields Brown or Blue: A Comment on the 2014 IPCC Report. *Economic Affairs* 34 (3): 399–401.

Seo, S. Niggol. 2014b. Evaluation of the Agro-Ecological Zone Methods for the Study of Climate Change with Micro Farming Decisions in Sub-Saharan Africa. *European Journal of Agronomy* 52 (January): 157–65.

Seo, S. Niggol. 2015. Modeling Farmer Adaptations to Climate Change in South America: A Micro-Behavioral Economic Perspective. *Environmental and Ecological Statistics* 23 (1): 1–21.

Seo, S. Niggol. 2020. *The Economics of Globally Shared and Public Goods*. Amsterdam, NL: Academic Press.

Seo, S. Niggol, and Robert Mendelsohn. 2008. Measuring Impacts and Adaptations to Climate Change: A Structural Ricardian Model of African Livestock Management. *Agricultural Economics* 38 (2): 151–65.

Shaw, M.R., E.S. Zavaleta, N.R. Chiariello, E.E. Cleland, H.A. Mooney, and C.B. Field. 2002. Grassland Responses to Global Environmental Changes Suppressed by Elevated CO_2. *Science* 298 (5600): 1987–90.

Wilder, L.I. 1932. *Little House on the Prairie*. New York, NY: Harper & Brothers.

Wired. 2021.This Burp-Catching Mask for Cows Could Slow Down Climate Change. Published on 1 January 2021. Accessed from https://www.wired.co.uk/article/cows-climate-change-methane-stop#.

World Rangeland Learning Experience (WRANGLE). 2020. *North American Short Grass Prairie*. Tucson, AZ: College of Agriculture and Life Sciences, University of Arizona.

World Wildlife Fund (WWF). 2020. *Northern South America—In Colombia and Venezuela: Tropical and Subtropical Grasslands, Savannas and Shrublands*. Gland, CH: WWF.

CHAPTER 7

Energy Revolutions: A Story of the Three Gorges Dam in China

7.1 ENERGY CONSUMPTIONS AND GLOBAL WARMING

The trend of global warming that we observe today is unavoidably tied to a sharp increase in energy consumption by humanity since the first oil-well was successfully drilled. The increase in energy consumption in turn was inevitable too. Consider this! The global population has grown by more than 7 times since the dawn of the twentieth century. Further, the global economic production has expanded by over 100 times during the same time period (Seo 2020).

Indeed, without the extravagant energy consumption of today, much of the anthropogenic and economic activities of today would be severely constrained, if not impossible. Energy is needed to provide electricity to homes and factories, to provide heating and air conditioning, to run automobiles and machines, to send satellites and spacecrafts. Energy is an essential input for the productions of foods as well as everyday consumer items.

Unfortunately, of about 10 giga tons of carbon (GtC) of global emissions annually (alternatively, about 36 $GtCO_2$), roughly 8 GtC (about 29 $GtCO_2$) is released annually from the energy generations and cement production only (Refer to Fig. 3.1, Chapter 3). The cement production is a highly energy intensive production process, so is customarily included in the carbon emission statistics (Ciais et al. 2013).

© The Author(s), under exclusive license to Springer Nature Switzerland AG 2021
S. N. Seo, *Climate Change and Economics*,
https://doi.org/10.1007/978-3-030-66680-4_7

At the center of the energy sector emissions of carbon lie the fossil fuels such as coal, oil, and natural gas. Coal-fired power plants, oil-fired power plants, and natural gas-fired power plants together provide about 90% of the electricity that is consumed by the citizens of the US (Muller et al. 2011). In addition, automobiles burning gasoline, natural gas, and LPG (Liquefied Petroleum Gas) emit a high volume of carbon annually.

Faced with the global warming challenge of the Planet, you might ask: Should the human society reduce the energy consumption to save Earth? A large reduction in energy consumption at the global level, however, seems almost infeasible, although an individual may conduct numerous energy-saving activities such as turning off lights, taking public transportations, and insulating homes better. Even with these energy-saving measures adopted enthusiastically, the global level emissions of carbon will continue to rise owing to an expansion of the economy as well as an increase in population.

A more pertinent question has been whether the world can switch to alternative energy sources and technologies that emit a smaller amount of carbon for the same amount of energy generated (Heal 2010). These alternative energy sources are referred to as a low-carbon energy source. Some of these are a zero-carbon energy source, which releases nearly no carbon from energy productions.

The low-carbon energy sources include solar radiation, wind energy (offshore and onshore), geothermal energy, bio fuels, and hydro power (IPCC 2011). For automobiles, electric vehicles are touted as a zero-carbon mobile transportation that may replace internal combustion engines or internal compression (diesel) engines that rely on fossil fuels.

For lighting, novel methods of lighting have emerged and been touted for high energy efficiency, that is, a larger amount of light produced per unit of electricity input. The novel technologies include the Compact Fluorescent Lamps (CFL) and the Light Emitting Diodes (LED). Against the conventional incandescent lightbulbs, the LED lamps are estimated to produce over 50 times the amount of light energy than that produced by the incandescent lightbulbs from the same amount of electricity input (Akasaki et al. 2014).

Of the range of alternative energy technologies, the hydroelectric power is the most well-known and accounts for about 80% of the total alternative energy generations in the US (USEIA 2020). This is interesting considering the heavy emphasis laid on solar energy, or wind energy to a smaller degree, by policy-makers. In some countries such as Brazil, the hydroelectric energy is a dominant form of energy generations, specifically, accounting for over two-thirds of the country's total annual energy generation.

This chapter will focus on the potential and new initiatives around the hydroelectric energy generation and the present author will come back in Chapter 8 to explain the other alternative energy generations, especially solar energy. There are a couple of reasons that the hydroelectric power should be of high pertinence to this book, one of which is the ongoing undertakings in many parts of the world of constructing a massive hydroelectric dam.

The mega hydro power projects that have received much attention include the Three Gorges Dam on the Yangtze River, China; the Itaipu Dam on the Parana River between Brazil and Paraguay; the Inga Falls Dam on the Congo River in Africa; and the Yarlung Tsangpo River Dam planned in the Tibetan Plateau. Of these mega dams, I will explain the generation capacities of the hydro powerplants and their potential as a low-carbon energy generation technology via an assiduous look at the Three Gorges Dam in China.

7.2   Fossil Fuels

To assess the value of a low-carbon and renewable energy technology including the hydroelectricity, we need to start with an understanding of fossil fuels. The fossil fuels are a basket of high-carbon as well as non-renewable energy sources. They emit a high amount of carbon when burned and they cannot be replenished once used up.

Specifically, the fossil fuels refer to the fuels such as coal, petroleum, and natural gas collectively. These fuels were formed from dead carbon-based animals and plants through their anaerobic decompositions underground or under the sea for millions of years. Dead bodies of tiny phytoplankton in the seas, one by one, accumulate on the sea floor for millions of years in an oxygen-free environment. The solid rock formed in this way of finely decomposed bodies is then cooked under the Earth surface to become hydrocarbons, that is, liquid carbon (Joskow 2013).

Why has the human society relied so much on the fossil fuels for obtaining energy, which after all has contributed immensely to the incredible economic prosperity since, to point one moment in history, the first

extraction of crude oil in Erie, Pennsylvania at the end of the nineteenth century? What makes them so valuable resources for the economic
well-being and livelihoods of people?

The fossil fuels are valued because they release energy, in the form
of heat, when they are burned. The thermal energy is then utilized to
produce electricity and for other valuable works such as heating homes.
The heat energy is also produced from burning woods or other materials, say, dried cow dungs. Donkeys, cows, camels, and horses also
provide energy to humankind, from which numerous difficult tasks can
be accomplished. Then, why do we rely on fossil fuels so heavily?

The answer lies in the high energy density of fossil fuels compared with
the other energy sources. Let's think about it this way: To move a train
from the Grand Central Station to the Penn Station in New York City,
you would need hundreds of donkeys to complete the work. The work,
however, can be completed by just a small bottle of gasoline or natural
gas!

The amount of heat energy produced from fossil fuels is measured by
the British Thermal Unit (BTU) or joule which is the unit of energy. One
BTU is about 1,055 joules. One BTU is the amount of heat needed to
raise the temperature of one pound of water by one degree in Fahrenheit.
The energy consumption and production are commonly expressed in the
BTU. In 2017, the world energy consumption was 582 quadrillion BTU
or 582 billion MMBTU (Million BTU) (USEIA 2016).

You may further ask: Why is there so much energy stored in fossil fuels?
It can be attributed to the photosynthesis of living organisms, which was
explained in detail in Chapter 3. You may recall that the photosynthesis
is the process of converting sunlight, enhanced by carbon dioxide and
water, to the energy needed for living organisms. In particular, the process
results in a carbohydrate, a compound of carbon and hydrogens, which
is sugar. At the same time, the process accumulates carbons in the bodies
of planktons, plants, and animals. When these organisms die, the carbons
are then accumulated on the sea floor to become fossil fuels after millions
of years (Ciais et al. 2013).

At the other end of the life cycle, when humans burn fossil fuels, that
is, coal, crude oil, and natural gas, for numerous purposes, carbons are
released again back into the atmosphere. Because of the large amount of
carbon emitted from the combustion of fossil fuels, they are referred to
as high-carbon energy sources or carbon intensive energy sources.

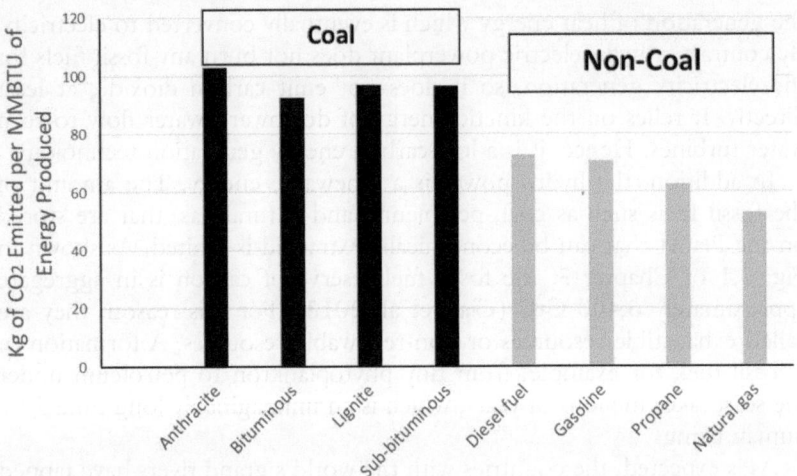

Fig. 7.1 Carbon intensity of fossil fuels

The amount of carbon dioxide released by a fuel for producing one MMBTU of thermal energy, which is the fuel's carbon intensity (per energy generated), is depicted in Fig. 7.1. The figure includes four types of coal: anthracite, bituminous, lignite, sub-bituminous. Along with gasoline, natural gas, and propane, diesel fuel and heating oil are also included. Diesel fuel can be either a fossil fuel or a renewable source depending upon its sources.

Against the anthracite coal which emits 102 kg of CO_2 per MMBTU of heat energy, as the figure shows, gasoline and natural gas are relatively lower-carbon fossil fuels. Gasoline emits 70 kg of CO_2 while natural gas emits 52 kg of CO_2 for the generation of the same amount of energy. Natural gas is a relatively cleaner fossil fuel, with about half of the amount of carbon dioxide emissions emitted by burning low-quality coals (USEIA 2020).

7.3 Hydroelectricity
as a Low-Carbon Renewable Energy

The large amount of carbon dioxide emissions is released as a byproduct of burning each of the above fossil fuels in Fig. 7.1, which is necessary for the generation of heat energy which is eventually converted to electricity. By contrast, a hydroelectric powerplant does not burn any fossil fuels for the electricity generation, so it does not emit carbon dioxide, at least directly. It relies on the kinetic energy of downward water flow to turn water turbines. Hence, it is a low-carbon energy generation technology.

In addition, the hydro power is a renewable energy. The amount of the fossil fuels such as coal, petroleum, and natural gas, that are stored on the Planet that can be economically extracted is limited. As shown in Fig. 3.1 of Chapter 3, the fossil fuel reserve of carbon is in aggregate approximately 6,400 GtC (Ciais et al. 2013). For this reason, they are called exhaustible resources or non-renewable resources. A formation of a fossil fuel, for example, from tiny phytoplankton to petroleum under the seas, takes millions of years, which is an unimaginably long time for a human being.

As is expected, the countries with the world's grand rivers have tapped into water resources to produce electricity in a way not relying on fossil fuels. Perhaps, the most famed one is the Hoover Dam in the US built during the Great Depression in the early twentieth century on the Colorado River. The world's biggest producer of hydroelectricity, as of 2020, is China, utilizing water flows from the mighty rivers such as the Yantze River and the Yellow River (IHA 2018). The Three Gorges Dam built on the Yantze River is the world's largest hydroelectricity dam, which will be the focus of the next section. China alone accounts for one-third of the world's total hydro power capacity, with an estimated 320 gigawatt (GW) of installed capacity.

The great river basins of Asia are depicted in Fig. 7.2 (FAO 2020). The Yantze River basin and the Huang He River (Yellow River) basin are in China. The Indus River basin and the Ganges-Brahmaputra River basin run through India and her Himalayan neighbors. The Mekong River runs through Vietnam and her neighbors while the Chao Phraya River through Thailand.

As per the reliance on hydro power, Brazil is the top hydroelectricity supplier. Brazil meets 75% [with a range from 62% to 82%] of its total energy demand by hydroelectric power (IHA 2018; World Bank 2019). The Itaipu Dam on the border of Brazil and Paraguay is one of the two world's largest dams in terms of electricity-generating capacity, along with the Three Gorges Dam. Canada is another top hydro power country. Roughly 56–63% of its total electricity production is accounted for by

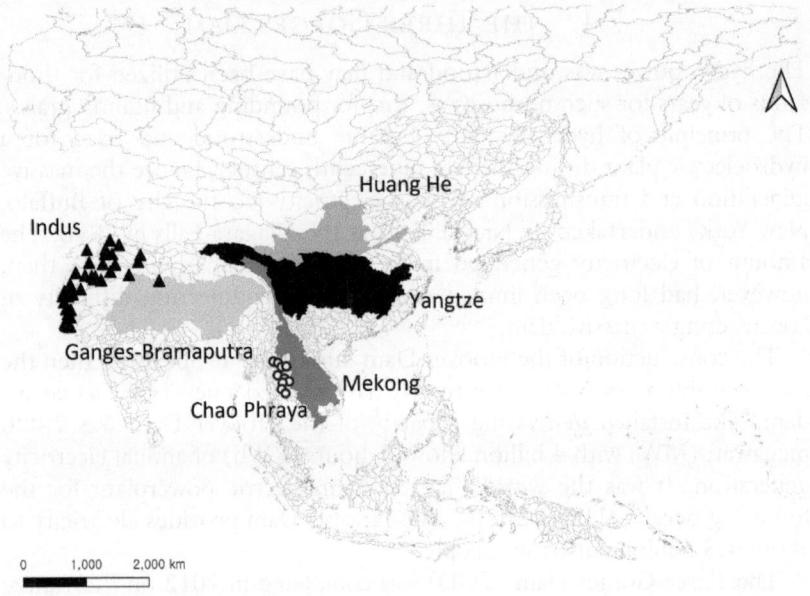

Fig. 7.2 Great river basins of Asia

hydroelectricity (IHA 2018; World Bank 2019). For reference, the US reliance on hydro power is about 6% of the total electricity production.

I said above that the hydroelectric power is a low-carbon energy generation system. Why is it not a zero-carbon energy generation technology if it relies on water flow alone to turn water turbines? This is due to an indirect emissions of carbon dioxide. The construction of a hydroelectric dam cannot be completed without making use of a large amount of cement/concrete and steel, whose productions are all highly carbon intensive. In addition, a production of essential machines such as water turbines cannot avoid releasing carbon dioxide. A diversion of the waterflow of the river for the dam construction, an essential construction process, has also significant carbon consequences by affecting soils and ecosystems. However, when the hydroelectric dam is built and starts generating electricity, there is little emissions of carbon dioxide per unit of electricity generated.

7.4   THE THREE GORGES DAM

The hydro power was understood and may have been utilized for thousands of years for such purposes as, *e.g.*, for pounding and hulling grains. The principle of hydroelectricity became understood and tried for a hydroelectric plant during the late nineteenth century, before the massive generation and transmission of hydroelectricity (to the city of Buffalo, New York) undertaken by Nicola Tesla at the Niagara Falls in 1896. The amount of electricity generated from numerous dams built since then, however, had long been limited owing to the engineering difficulty of constructing a massive dam.

The construction of the Hoover Dam during the 1930s using then the inconceivable amount of concrete may have opened a new era of a massive dam. The installed generating capacity of the Hoover Dam was 2,080 megawatt (MW) with 4 billion kilowatt-hours (KWh) of annual electricity generation. It was the world's largest hydroelectric powerplant for the following decade (USBR 2020). The Hoover Dam provides electricity to about 1.3 million American people.

The Three Gorges Dam (TGD) was completed in 2012 on the Yantze River in China to become the world's largest hydroelectric dam. You can confirm the location of the Yantze River basin in Fig. 7.2. The massive scale of the TGD can be described in comparison with the scale of the Hoover Dam. It has the generating capacity of 22,500 MW, around 10 times larger than that of the Hoover Dam. It generates about 100 billion kilowatt-hours of electricity annually, about 25 times that of the Hoover Dam.

As per the structure, the Three Gorges Dam is 2.3 km long and rises as high as 181 meters from the bottom (USGS 2020). It has 34 generators: 32 main generators and 2 plant power generators. Each main generator has a generation capacity of 700 MW.

The TGD generated about 100 billion KWh of electricity in 2018, or equivalently 100 terawatt hours (TWh) of electricity. Taking into consideration the per capita electricity consumption in China which is about 4 MWh/year, the TGD alone can provide electricity to 25 million people in China (World Bank 2019).

For comparison, the Itaipu Dam in Brazil had been the world's largest hydroelectric dam since 1984 until the completion of the Three Gorges Dam. The Itaipu Dam has the generating capacity of 12,600 MW.

However, the Itaipu Dam generates as much electricity as the TGD annually, about 105 TWh, because of the seasonal variations in water flow in Yantze River. With the per capita electricity consumption of 2.6 MWh per year in Brazil, the Itaipu Dam alone could provide electricity to about 40 million Brazilians.

At the present moment, another mega hydroelectric powerplant project is being undertaken in Sub-Saharan Africa. The Grand Inga Dam (GID) that began to be built at the Inga Falls on the Congo River in central Africa would become the largest dam once completed with the generating capacity of 40,000 MW (Pearce 2013). With an assumption of 200 TWh annual production, the GID alone could provide electricity to 1 billion Africans with the current annual electricity consumption in 2018 of about 200 KWh per capita in Africa which is only 1/20th of China's. For reference, the total African population is 1.26 billion.

A historical growth of the hydroelectricity generation is drawn in Fig. 7.3 which summarizes the installed hydroelectric capacity of the six historical dams in the major hydroelectricity generating nations: Hoover Dam (1939, US), Grand Coulee Dam (1974, US), Robert-Bourassa Dam (1981, Canada), Itaipu Dam (1984, Brazil and Paraguay), Three Gorges

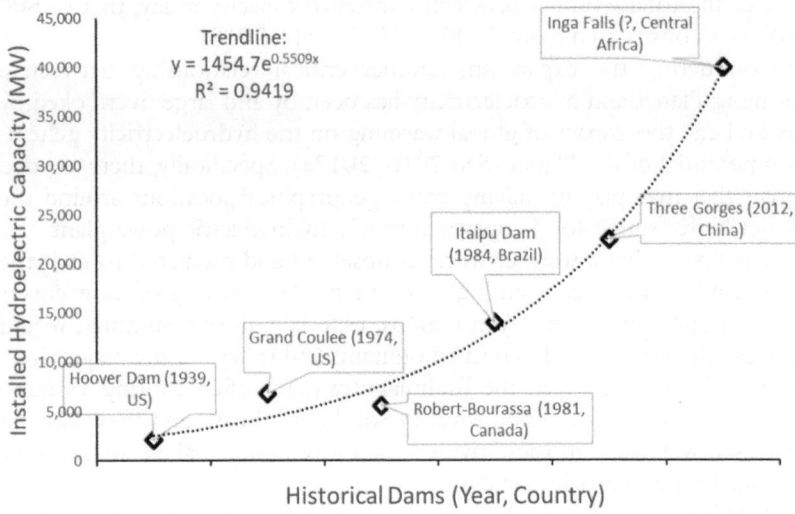

Fig. 7.3 The installed hydroelectric generation capacity for historical dams

Dam (China), and Inga Falls Dam (Central Africa). The exponential trendline with a very high degree of fit shows that the dam construction has been expanding in an exponential manner.

The International Hydropower Association's status report in 2018 reports that the worldwide installed generation capacity is 1,267 GW and the annual hydroelectric generation is about 4,185 TWh in 2017. Of the total capacity, China accounts for 319 GW, the US accounts for 101 GW, Brazil 92 GW, and Canada 79 (IHA 2018). For reference, the total electricity generation of the world in 2016 was slightly over 25,000 TWh, which means that the hydroelectricity accounted for about 17% of the total electricity generation worldwide (World Bank 2019).

7.5   Economics I: Go for Hydro Under a Warmer Planet?

The discussions up to now indicate that the hydroelectricity is an excellent alternative technology for producing electricity if we hope to reduce the energy-related emissions of carbon dioxide. This point is reflected in the rapid expansion of hydroelectricity generations worldwide in the past two decades described above. According to the IHA report cited above, about 45% of the total world hydroelectric installed capacity today, that is, 500 GW, was constructed from 2000 to 2017 (IHA 2018).

Considering the expansion, another critical relationship between a warming Planet and hydroelectricity has been by and large overlooked by researchers: the impact of global warming on the hydroelectricity generation potential of the Planet (Seo 2016, 2017a). Specifically, there are two forces that may play in making many geographical locations around the Planet better suited for a construction of a hydroelectric powerplant.

The first is that a warmer world is observed and predicted to melt the snow covers and ice caps on high mountains. The melting of snow covers in Tibet and other Himalayan regions, especially during summer, would increase the waterflow down the mountains and valleys to the major rivers such as the Yantze River, the Brahmaputra (also called Yarlung Tsangpo in Tibet) River, and Mekong River in South, Southeast, and East Asia. An increased waterflow in these rivers would give policy-makers an incentive to construct a hydroelectric dam.

A hydroelectric dam built along these rivers can serve a more important public policy purpose other than climate change mitigation. That is, it would store the downstream waterflow from the Himalayan mountains and supply it for agricultural consumptions during the growing seasons as well as household consumptions. Therefore, it could play an important role in food security and water security of the nations with the world's high mountains.

The first force may explain the expansion of hydroelectric power generations in recent decades in China, as described above, and other Southeast Asian countries. In these countries, a hydroelectricity expansion is recognized as one method to fulfill their commitments made to the Paris Agreement, that is, through an increase in the percentage of renewable energy in the portfolio of total energy production (UNFCCC 2015; Seo 2017b).

The second force is more difficult to verify because it concerns the predictions of regional precipitations in the future. A host of sophisticated global climate models predict that a higher global temperature in tandem with a higher ocean temperature will increase the amount of precipitation in many parts of the Planet. The simplest mechanism in biogeochemistry explains this, although many regional factors can offset this: A higher temperature at the Earth surfaces increases the evaporation of water from land and ocean surfaces, which becomes a source of increased cloud formations and ultimately increased rainfall (IPCC 2014). An increased precipitation means a larger waterflow at some rivers, which can be tapped for a hydroelectricity generation at a strategic location.

The two forces, in addition to it being one of the low-carbon energy technologies, could make the hydroelectricity one of the winners in the race for the future energy technologies in a warmer Earth. This well-known centuries-old technology, if the predictions made in this section should turn out to be true, will be reborn as a future technology for the Planet especially at selective favorable geographic locations.

7.6 ECONOMICS II: IS HYDROELECTRICITY TOO COSTLY LIKE SOLAR ENERGY?

The list of low-carbon and/or renewable energy generation technologies is fairly long. It ranges from a solar Photovoltaic (PV) system, onshore/offshore wind energy, geothermal energy, biofuels, biomass, wave energy, nuclear fission, nuclear fusion, and a hydroelectric system. A subset of the family of low-carbon technologies is an advanced system of fossil-fuel-based energy generation technologies, for example, ones with a carbon-capture-storage capacity embedded (IPCC 2011).

Any attempt to transform the nation's energy system from a high-carbon system to a low-carbon system will and has faced the same dilemma: a higher cost of energy from the latter. The low-carbon energy generation technologies are markedly more expansive to build and operate than the high-carbon energy generation technologies. Concretely, the costs of the renewable and/or low-carbon energy technologies are 50–100% higher per unit of energy provided than the costs of the fossil fuel-based technologies (Heal 2010). As such, an individual household or business who wishes to make a switch to a low-carbon energy technology must be ready to pay the additional cost, which will certainly discourage some households and businesses from making an actual switch.

Will the hydroelectricity generation face the same hurdle of too high cost of electricity against the presently common cost of electricity generation from burning fossil fuels? To explore this question, I show in Fig. 7.4 the relative costs of alternative electricity generation technologies with reference to the conventional coal-fired electricity generation technology, based on the estimates from the Energy Information Administration (EIA) in the US government (USEIA 2020). The data shown

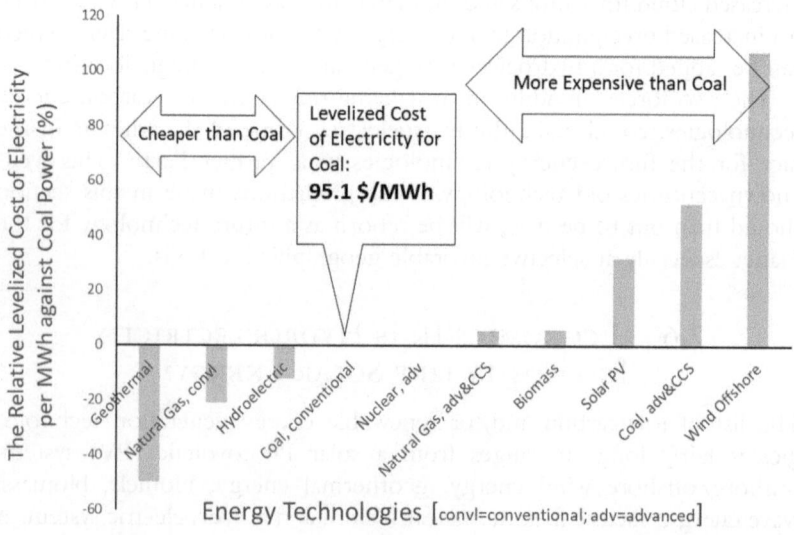

Fig. 7.4 Relative costs of alternative electricity generations against the coal-fired power generation

here are from the Annual Energy Outlook (AEO) 2015. Readers can refer to the AEO 2020 for the recent estimates for the alternative power generation plants entering in 2025 and 2040.

The levelized cost of the electricity from the coal-fired power plant is US$95.1 per MWh of electricity supplied. The levelized cost is defined as the average cost of supplying 1 MWh of electricity during the entire lifespan of a powerplant, as such, it includes the cost of construction, operation, and maintenance. The figure confirms that some low-carbon energy generations are by far more costly. The solar PV system is 32% more costly than the coal-fired electricity generation system; the coal-fired power plant with a carbon-capture-storage facility is 52% more costly; and the offshore wind system is 107% more costly than the conventional coal-fired system. Notably, the AEO 2020 predicts that the LCOE of the solar PV will fall rapidly and far below the energy cost from the coal-fired power generation by 2040 (USEIA 2020).

The figure also offers an answer to the top question of this section: The cost of hydroelectricity is, on the contrary to our initial concern, lower than the cost of electricity from the conventional coal-fired power generation. It is about 12% lower, with the levelized cost of US$83.5 per MWh of electricity generated.

The other less costly options for a low-carbon energy generation shown in the figure are geothermal and natural gas. The geothermal energy can be generated with only half the cost of the coal-fired power generation, but has a severe scale problem. In other words, a geothermal electricity generation is suitable at ideal locations but difficult to scale up for a national energy supply (USEIA 2016).

The natural gas-fired electricity generation is estimated to cost 20% less than the cost of the coal-fired system, which makes it an attractive alternative to the coal-fired power generation. As shown in Fig. 7.1, the natural gas-fired system emits only half the carbon dioxide emissions of the coal-fired system. The discovery and production of natural gas, more often called shale gas, expanded vastly in the past decade owing to the advance of a hydraulic fracture technique (Joskow 2013). The concerns on the fracking as well as the ways to address them effectively have simultaneously gained recognition: to name some, methane leaks, geologic stability, and water quality.

Note also in Fig. 7.4 that the nuclear power system can produce electricity at about the same cost as that from the coal-fired powerplant, which makes it another attractive energy generation option for humanity. It is often viewed as a zero-carbon energy option. The major hurdle to expand nuclear (fission) energy is a safe long-term storage of nuclear wastes (USDOE 2020). Nuclear fusion is another zero-carbon energy generation technology, which I will describe in detail in Chapter 8 as one of the backstop technologies.

7.7   LIVELIHOODS AND ECOSYSTEMS

Up to now, I have painted a favorable picture of the hydro power as an energy option in the age of global warming. Now it is opportune time to note and recognize the reality that the hydropower generation is a politically sensitive issue. To put it more directly, a hydroelectric dam, once planned and built, will affect the livelihoods of millions of people around the dam, especially downstream villages of the dam.

To build a massive dam such as the Three Gorges Dam, millions of people in downstream villages and the surrounding areas should be relocated first. The river must be diverted during the construction of the dam, for which large multiple diversion channels have to be built. The effects on the villages are severe, to say the least. Such a large-scale population relocation would be nearly impossible in most democratic nations. With the relocation, it would be unavoidable to destroy or damage historically, culturally, and religiously important sites to the country for a long time. Considering this, a successful dam construction depends critically on choosing a "favorable" location.

The political sensitivity does not stop there. The dam construction may also severely affect the ecological systems of the neighboring and downstream areas. It is nearly certain that the dam construction will destroy many of the local ecological systems, while some of which will be perceived by the general public to be of crucial importance to them. Sometimes, the public's grave concern falls on as a "minor being" as a frog community. It is possible to argue that there will emerge a new ecological system once the dam is completed and the new ecology will be as good as the old ecology. Nonetheless, some of the old ecology may be lost "forever" indeed.

To sum up: Faced with the challenges of global warming, we must therefore make the following choice: Should a massive hydroelectric dam be built for the sake of the Planet at the expense of livelihoods of many and ecosystems? Or should the livelihoods of the villages and local ecological systems be preserved forever and for future generations even at the expense of an important global climate strategy? The answer to these questions will vary from one country to another. The answer is also not dichotomous since there are other low-carbon energy options than hydroelectricity.

7.8 Chapter Highlights

- The global energy production (cement production included) which relies dominantly on fossil fuels accounts for 80% of global annual carbon emissions.
- This chapter explains the fossil fuel-based electricity generations such as coal, petroleum, and natural gas, which are then contrasted with low-carbon energy technologies such as solar PV, offshore wind turbines, nuclear energy, and geothermal energy.
- Unlike many other low-carbon energy technologies which are still too costly to produce, a hydroelectric powerplant can produce electricity at a lower cost than a coal-fired powerplant.
- Further, global warming may aid the developments of the hydro power via increased melting of snow covers and ice caps from the world's high mountains as well as an increase of precipitation in some regions.
- The global hydroelectricity generation has doubled in the past two decades and the engineering advances have made it possible to build a massive dam. The Three Gorges dam built in Yantze River in China has the installed generation capacity that is about 10 times larger than that of the Hoover dam.
- The other massive dams include the Itaipu dam at the Brazil-Paraguay border, the Inga Falls dam being constructed in central Africa, and the Yarlung Tsangpo River dam being constructed in Tibet. The hydroelectricity accounts for roughly 75% of electricity consumption in Brazil and roughly 60% in Canada.

References

Akasaki, Isamu, Hiroshi Amano, and Shuji Nakamura. 2014. *Blue LEDs: Filling the World with New Light*. Nobel Prize Lecture. Stockholm, SE: The Nobel Foundation. Accessed from http://www.nobelprize.org/nobel_prizes/physics/laureates/2014/popular-physicsprize2014.pdf.

Ciais, P., C. Sabine, G. Bala, L. Bopp, V. Brovkin, J. Canadell, A. Chhabra, R. DeFries, J. Galloway, M. Heimann, C. Jones, C. Le Quéré, R.B. Myneni, S. Piao, and P. Thornton. 2013. Carbon and Other Biogeochemical Cycles. In *Climate Change 2013: The Physical Science Basis.* Cambridge: Cambridge University Press.

Food and Agriculture Organization (FAO). 2020. *Hydrological Basins in Southeast Asia—Derived from HydroSHEDS.* Rome, IT: FAO-Geo Network.

Heal, Geoffrey. 2010. Reflections—The Economics of Renewable Energy in the United States. *Review of Environmental Economics and Policy* 4 (1): 139–54.

International Hydropower Association (IHA). 2018. *Hydropower Status Report.* London, UK: IHA.

Intergovernmental Panel on Climate Change (IPCC). 2011. *Special Report on Renewable Energy Sources and Climate Change Mitigation.* Cambridge: Cambridge University Press.

Intergovernmental Panel on Climate Change (IPCC). 2014. *Climate Change 2014: The Physical Science Basis, The Fifth Assessment Report of the IPCC.* Cambridge: Cambridge University Press.

Joskow, Paul L. 2013. Natural Gas: From Shortages to Abundance in the United States. *American Economic Review* 103 (3): 338–43.

Muller, N.Z., R. Mendelsohn, and W. Nordhaus. 2011. Environmental Accounting for Pollution in the United States Economy. *American Economic Review* 101: 1649–75.

Pearce, Fred. 2013. *Will Huge New Hydro Projects Bring Power to Africa's People? Yale Environment 360.* New Haven, CT.

Seo, S. Niggol. 2016. A Theory of Global Public Goods and Their Provisions. *Journal of Public Affairs* 16 (4): 394–405.

Seo, S. Niggol. 2017a. *The Behavioral Economics of Climate Change: Adaptation Behaviors, Global Public Goods, Breakthrough Technologies, and Policy-making.* Amsterdam, NL: Academic Press.

Seo, S. Niggol. 2017b. Beyond the Paris Agreement: Climate Change Policy Negotiations and Future Directions. *Regional Science Policy & Practice* 9 (2): 121–40.

Seo, S. Niggol. 2020. *The Economics of Globally Shared and Public Goods.* Amsterdam, NL: Academic Press.

United Nations Framework Convention on Climate Change (UNFCCC). 2015. *The Paris Agreement. Conference Of the Parties (COP) 21.* New York: UNFCCC.

United States Bureau of Reclamation (USBR). 2020. *Hoover Dam.* Washington, DC: USBR. Accessed from https://www.usbr.gov/lc/hooverdam/faqs/pow erfaq.html.

United States Department of Energy (US DOE). 2020. *Waste Isolation Pilot Plant (WIPP)*. Washington, DC: US DOE. Accessed from http://www.wipp.energy.gov/.

United States Energy Information Administration (US EIA). 2016. *Energy Explained*. Washington, DC: EIA.

United States Energy Information Administration (US EIA). 2020. *Annual Energy Outlook 2015 & Annual Energy Outlook 2020*. Washington, DC: US EIA, Department of Energy.

United States Geological Survey (USGS). 2020. *Three Gorges Dam: The World's Largest Hydroelectric Plant*. Reston, VA: USGS.

World Bank. 2019. *World Development Indicators*. Washington, DC: The World Bank.

CHAPTER 8

Backstop Technologies: A Story of a Humble Greenhouse with Surprises

8.1 GREENHOUSE EFFECTS

When it comes to a greenhouse, you are likely already firmly wired to the term "greenhouse effects" through your school education and media contacts up to this point in your life. The term is one of the landmarks in the science of climate change, as such, should be placed at a central place in any scientific report and policy forum. So, I will have to begin this chapter with an explanation of the popularly known term and then move on to clarify the greenhouse whose story I have a plan to tell you in this chapter.

It was not the case for the present author, being born in a remote rural village in so-called Far East and having grown up there until the 4th grade in the primary school. That is to say, the term "greenhouse" meant something else quite personal to the present author. As a kid in a rural farming family, one of the responsibilities of mine was to take care of the greenhouses in the events of strong storms and heavy rains, in which cases the greenhouses should be rolled down and fastened manually in time. The stories to be told in this chapter have much to do with this early experience of the present author.

Let me start with the two terms, greenhouse effects and greenhouse gases, in this section, then explain a farm greenhouse in Sect. 8.2 and greenhouse technologies in Sect. 8.3. The greenhouse effect refers to the Planet-warming effect of various chemical compounds that are in the

© The Author(s), under exclusive license to Springer Nature Switzerland AG 2021
S. N. Seo, *Climate Change and Economics*,
https://doi.org/10.1007/978-3-030-66680-4_8

Earth's atmosphere (UNFCCC 1992). The most prominent of them is carbon dioxide (CO_2). A large increase in the amount of carbon dioxide in the atmosphere is compared to a formation of a transparent blanket in the sky whose primary function is to heat the Planet up by passing the incoming sunlight through to Earth surface but by blocking the reflected sunlight from going out to space. The reflected sunlight, more formally solar radiation, by the Earth surface is a long-wave radiation referred to as the infrared radiation.

The greenhouse effect is measured in scientific unit as Radiative Forcing (RF). The RF is roughly the difference (imbalance) between the incoming solar energy and the outgoing solar energy (Le Treut et al. 2007). The amount of solar radiation is measured by Watts per square meter (W/m^2). The Watts is the amount of energy needed to produce a certain amount of light. The higher the Watts, the brighter the light. If you have a lightbulb in your bedroom, it will likely be lit by a 60 watts incandescent bulb or a 15 watts LED lamp. The radiative forcing of carbon dioxide is defined as the additional radiation incident on Earth because of the greenhouse-like blanket formed in the atmosphere by carbon dioxide.

Besides carbon dioxide (CO_2), other chemical compounds have similarly Planet-warming effects when they are released into the atmosphere. Such chemical compounds are nitrous oxides (N_2O), methane (CH_4), fluorinated gases such as hydrofluorocarbons (HFCs), and water vapor. These gases are collectively referred to as greenhouse gases (GHGs).

It should be noted that there are other types of gases that have the opposite effect, that is, the Planet-cooling effect when they are released in the air. The aerosols are fine particles or tiny sulfate droplets suspended in the air, whose major function is to increase the reflectivity of incoming solar radiation, thereby canceling out some of the Planet-warming effects caused by the aforementioned greenhouse gases.

What is the magnitude of radiative forcing that is occurring at the present time due to the increase in carbon dioxide in the atmosphere from the pre-industrial level of CO_2? Scientists estimate it to be 1.6 W/m^2. The radiative forcings from the other greenhouse gases are also measured and made available by climate scientists. The cooling effect of the aerosols is expressed as a negative radiative forcing.

One of the central questions in climate science is how large the magnitude of the RF will be from a doubling of carbon dioxide concentration in the global atmosphere from the pre-industrial CO_2 concentration. Scientists estimate it to be $+3.7 \, W/m^2$.

The larger the magnitude of radiative forcing of the greenhouse gases, the larger the degree of the Planet's warming. More concretely, the magnitude of radiative forcing caused by the carbon dioxide doubling will determine, through the feedback loops, the magnitude of global warming under the doubled carbon dioxide environment. This is expressed by climate scientists as the equilibrium climate sensitivity which is now estimated to be 3 degrees in Celsius. This is called the best guess estimate. The range of the equilibrium climate sensitivity, "adopted" by the Intergovernmental Panel, is quite wide: from 1 °C to 4.5 °C (IPCC 2013).

8.2 A GREENHOUSE

Let us get back to the greenhouse itself before any global warming connotations attached to it. A greenhouse is a structure, often house-like or tunnel-like, that is designed to control the climate inside the structure for enhanced economic activities inside it. The climate variables that are controlled are mainly temperature, moisture, and solar radiation. This humble structure has been one of the most popular inventions that affected the agricultural and horticultural producers throughout history (Muijzenberg 1980).

A careful examination of the above description of the greenhouse will reveal to you that the term "greenhouse effect" places a one-sided emphasis on the temperature-raising capacity of the greenhouse. It will also reveal that the other aspect of the greenhouse structure is overlooked, that is, its essential role of enhancing economic productions.

Why bother this? You may ask. As will be elucidated in this chapter, the latter aspect of the greenhouse is turning out to be a pivotal concept in designing the efficient solutions to the problems of climate change (Seo 2020). In this concept, the greenhouse is a metaphor for the basket of breakthrough technologies that help control the climate system and simultaneously help improve economic productivities.

Let me clarify this point further. A greenhouse is one of the most primitive scientific inventions by humanity whose main capacity is to control the climate conditions inside the structure for the purpose of enhancing the outputs of specialty crops and horticultural products. There are many types of a greenhouse, also sometimes called a glasshouse. Across them all, the key structural feature is that it is covered by plastic (in polyethylene film) or glass, which makes it possible to pass sunlight through but block the outgoing reflected solar radiation. Modern commercial greenhouses are high tech production facilities in which a computer moderates an optimal climate condition for plants' growth (GLASE 2019).

The economic importance of a greenhouse for the agricultural communities is that it enables them to produce vegetables, specialty crops, and horticultural products even during the non-growing seasons of the year, say, the winter season. Farmers can keep producing and selling greenhouse products during the off-harvesting seasons, thereby earning additional incomes. It enables them to better cope with the natural farming cycle. In addition, it improves consumers' welfare because greenhouse products can be delivered to them year-round.

Who invented a greenhouse? A simple structure that is built to grow vegetables and flowers during the wintertime may be traced back to the empires thousands of years ago in Europe and Asia. A construction of the first modern practical greenhouse is sometimes credited to French botanist C.L Bonaparte who built a practical greenhouse in Leiden in the Netherlands during the 1800s. He built the structure to grow medicinal tropical plants, which became popular, quickly spreading to the gardens of the rich as well as the universities in Europe. They were called orangeries, pineries, etc., referring to the plants grown inside the structure.

The first description of a greenhouse in the historical records may be found in an East Asian annals. In the history book of Korea, there appears a detailed description of a greenhouse-like structure that were built during the 1450s. The royal history book gives a detailed instruction about how to construct a chamber in order to grow vegetables during the winter time. According to this, the chamber's inner space is heated through an *ondol* system which is a traditional heating system unique to Korea in which the heat is passed through the heat channels of stones and transferred through mud floors of the chamber. In addition, the chamber is built to have windows made of *Hanji*, a transparent paper, so that the sunlight passes through (Yoon and Woudstra 2007). However, it did not have a plastic cover or a glass cover common in the modern greenhouses. This type of structure was a common house structure in Korea and perhaps Japan and China also, not specifically intended as a greenhouse.

The greenhouse structure that we come across today in any rural area of the Planet came to fruition during the 1960s through the availability of polyethylene film, which is the most common plastic today. By then, the techniques of steel tubing and aluminum extrusions came to be used for a greenhouse construction (Muijzenberg 1980).

Another key innovation in the history of the greenhouse was made during the latter half of the twentieth century. As explained in Chapter 3, an increase in carbon dioxide concentration enhances the photosynthesis of crops and plants. The innovative greenhouse was first built in the Netherlands, which has the capacity of carbon dioxide enrichment for the purpose of increasing plant growth (Wittwer and Robb 1964).

Today, you will notice greenhouses as the most salient man-made structure in any rural areas of the world that you drive through, along with cowsheds in some areas. According to Erwin Muijzenberg, the first director of the Institute of Horticultural Engineering at Wageningen University, a quarter of the greenhouses in the world were found in the Netherlands during his time (Muijzenberg 1980). Today, the Netherlands, a tiny country in Europe whose land size is only 1/270th of the US, feeds the world, as the number 2 exporter of food in the world after the US, by resorting to high value farming accomplished through an array of a gargantuan greenhouse complex (Viviano 2017).

8.3   Greenhouse Technologies

The greenhouse, as explained in the above, can be interpreted as a metaphor for all the technologies whose primary function is a climate control and whose primary purpose is to gain economic benefits. In this context, the present author recently referred to some technologies with such characteristics as a greenhouse technology (Seo 2020).

What are such technologies, specifically? The above-cited book clarified three types of greenhouse technologies. The first type is a set of technologies that reflects the sunlight directly to cool Earth (NRC 2015a). The second type is a set of technologies that captures carbon dioxide directly from the atmosphere, thereby reducing the Plant-heating effect (NRC 2015b). The third type is a family of technologies that aims at supplying the energy to humanity without carbon dioxide emissions, specifically, the zero-carbon energy technologies discussed in the preceding chapter (ITER 2015).

Of the three types, the term a greenhouse technology is perhaps most aptly applied to the first type because of their similarity in appearances.

The core of the first type technology is a creation of a layer in the atmosphere whose function is to increase the solar radiation reflectivity of Earth, which then cools the Planet. The layer can be an aerosol—fine dust—layer or a cloud layer. The basket of technologies of the first type is often referred to as an Albedo modification (AM) technique. Albedo here is the reflectivity (NRC 2015a). It is also referred to as a solar radiation management (SRM) technique.

As far as their appearances are concerned, like the farm greenhouses, the Albedo modification technology creates a polyethylene-film-like layer in the air whose function is to alter the reflectivity of the Earth's atmosphere of solar radiation. The farm greenhouse can either increase or decrease the temperature within it while the SRM layer is intended to decrease the temperature underneath it.

The technologies of the second type are the ones that capture and remove carbon dioxide from the air (NRC 2015b). The most drastic one of them is a direct air capture and storage (DACS) in an open air. The less drastic one is a capture and storage facility embedded at a fossil-fuel-fired powerplant which was introduced in Chapter 7 as one component of energy revolutions. The DACS technology has been tested successfully at a local scale, but has not been tested at as sufficiently large a scale as can influence the Earth's climate or a nation's climate.

Once again, the DACS technology, by removing carbon dioxide from the air, aims at cooling the Planet by reducing the Planet-heating effect of carbon dioxide. The captured carbon dioxide can be stored for a long time and reused at a later time for, *e.g.*, enhancing plants' growth at the farm greenhouses (Wittwer and Robb 1964).

The greenhouse technologies of the third type refer to a radical technological transformation in the energy sector to a zero-carbon energy technology. The most discussed is a nuclear fusion technology, followed by a nuclear fission technology with a safe storage solution. The solar photovoltaic (PV) energy is also seen as a zero-carbon technology, but it may be insufficient to provide the energy needed by the global community by itself without another revolution in electricity storage. Further, the solar PV technology requires a large land area on Earth for an expansive solar array. But, a more efficient solar cell, including perovskite, may be able to reduce the cost of solar energy and increase the solar conversion rate significantly (NREL 2020).

The nuclear fusion energy has been touted for over half a century as the ultimate energy technology for humanity. However, the high expectation has not materialized in real life up until now. Unlike the conventional nuclear powerplant relying on nuclear fission, a nuclear fusion technology replicates a nuclear fusion process of the Sun in which hydrogen (H) atoms combine to form helium (He) gases, which gives off the radiant energy of the Sun (ITER 2015). For your reference, the Sun's core temperature reaches 15 million °C while the plasma temperature at a nuclear fusion plant reaches 150 million °C, which is perhaps the hell's temperature.

Despite the "simple" physics of nuclear fusion, engineering challenges to build an effective nuclear fusion plant have turned out to be massive. The ongoing experiments are found in the National Ignition Facility (NIF) at the Lawrence Livermore National Laboratory (LLNL), a Federal research facility located at the University of California at Berkeley and the ITER project in southern France. For the latter, 35 nations participate to build a Tokamak, a giant nuclear fusion machine (LLNL 2015; ITER 2015).

The engineering challenge can be explained by the following statistics. Up to now, the most efficient fusion machine in the world produces an output of 16 MW thermal fusion power for each 24 MW input heating power, the ratio of only 0.67. This ratio is the world record but below the plasma energy breakeven point ($Q = 1$). The ITER Tokamak is scheduled to open in 2025, so the Q efficiency improvement has yet to be seen. The ITER is designed to produce a tenfold return on energy ($Q = 10$), or 500 MW of fusion power from 50 MW of input heating power. Even with the target Q efficiency, the Tokamak would fall far short of the generating capacity of the Three Gorges hydroelectric dam which stands at 22,500 MW.

Considering these challenges, the nuclear fission energy is a proven technology with the above-described engineering challenges largely overcome as of today. In fact, more than a dozen countries already satisfy more than 30% of their energy consumption needs through the nuclear fission energy. To enumerate these countries by the nuclear energy dependence rate: France (76.9%), Slovakia (56.8%), Hungary (53.6%), Ukraine (49.4%), Belgium (47.5%), Sweden (41.5%), Switzerland (37.9%), Slovenia (37.2%), Czech Republic (35.8%), Finland (34.6%), Bulgaria (31.8%), Armenia (30.7%), South Korea (30.4%). The largest producer of the nuclear energy is the US which produces 8% of the country's total annual energy consumption by nuclear fission powerplants (NEI 2016).

8.4   ECONOMICS OF GREENHOUSE TECHNOLOGIES

In the previous section, I referred to the nuclear fusion energy as the ultimate energy, the term often used by others too. Similarly, we can refer to the greenhouse technologies of the first and second types as the ultimate technologies when it comes to controlling the Earth's climate system.

A formal scientific terminology for the ultimate technology, which best captures the core idea, is a backstop technology. Coined in the energy economics literature more than half a century ago, the essence of the term is that the global community will make a transition from one energy source to another, owing to their limited stocks and gradually rising prices over time due to ever diminishing stocks, and ultimately to the backstop energy technology. The backstop technology, say, a nuclear fusion technology, will provide humanity with the energy "indefinitely" and at a constant price (Nordhaus 1973).

The first and foremost economic question is therefore "at what price?". The present author described the backstop technologies in the previous section without pointing to their backstop costs/prices. The backstop cost is a pivotal factor because if the ultimate energy, say, a nuclear fusion or a solar PV, were to be too costly against the currently dominant technologies, say, a coal-fired or a natural gas-fired power generation, it would not be adopted as a replacement energy now and for a long time henceforth by most energy consumers.

Therefore, the cost of a backstop technology should be measured against the costs of fossil fuel-based energy generations as well as the costs of other low-carbon renewable energy generations. The comparisons would offer a key indicator on whether and when the backstop energy would replace the aforementioned transitional energies.

From another angle, the cost of the backstop technology is a critical concept in an economic analysis of climate change. If the Planet were to switch to the backstop technology as an ultimate solution for containing a dangerous change in the climate system, the total cost of the global community's efforts to contain the climatic change could not be greater the cost of the backstop technology. In other words, the backstop cost sets the upper bound to the Planetary cost of addressing the dangerous climate change. For the ease of comparison by interested parties, both costs are routinely expressed in the literature as US dollars per ton of carbon (or carbon dioxide) mitigated or removed.

Further elaborated, if we have a reliable estimate of the backstop cost, the world community will be able to avoid dwelling on or proceeding with the technologies and/or remedies that are way too more costly than what the backstop calls for. This brings us to the estimates of the backstop cost. Summarized in Table 8.1 are the cost estimates of the three type greenhouse technologies put forth by the National Research Council (NRC)'s two major reports (NRC 2015a, b).

The costs of the two Type I technologies are, according to the NRC assessments, by far lower than the cost of the conventional decarbonization of the economy, that is, regulating carbon emissions from economic activities. The two technologies are: a stratospheric aerosol layer and a marine cloud brightening. The latter is a technique of increasing white clouds to reflect solar radiation. They are estimated to cost only 1/10th of the economic cost of decarbonization.

Table 8.1 The costs of backstop technologies

Backstop types	Specific technologies	Cost estimates	Expert judgments on adoption possibility
Type I	A stratospheric aerosol layer	An order of magnitude smaller than the cost of decarbonizations to offset anthropogenic CO_2 increases (NRC 2015a)	Feasible but should not be deployed at this time, owing to risks (NRC 2015a)
	Marine cloud brightening	An order of magnitude smaller than the cost of decarbonizations to offset anthropogenic CO_2 increases than optimal mitigations (NRC 2015a)	Feasible but should not be deployed at this time, owing to risks (NRC 2015a)
Type II	Direct air capture and sequestration	380–600 US\$/tCO_2 (NRC 2015b)	An immature technology with only laboratory-scale experiments (NRC 2015b)
	Ocean iron fertilization		Too high risks (NRC 2015b)
Type III	Nuclear fusion	Currently, the input energy is larger than the output energy ($Q < 1$). The ITER is designed to produce a ten-fold return on energy ($Q = 10$) (ITER 2015)	Inefficient and not practical at present

As for the Type II backstop technologies, their costs are much higher than the cost of the conventional decarbonization. Deployed today, the direct-air-capture-storage technology is estimated to cost about US$380 to US$600 per ton of carbon dioxide. Converted to carbon unit, the cost is equivalent to US$1,400 to US$2,200 per ton of carbon. For your reference, an economically efficient management of the climate change problem would put the carbon price at about 250$ per ton of carbon at the end of the twenty-first century (Nordhaus 2013). For another reference, the social cost of carbon today is in the range of about 150$ per ton of carbon according to the survey of all economic estimates (Tol 2009).

The table also lists another Type II backstop technology not mentioned up until now in this book, the ocean iron fertilization. This is a "simple" technology of adding iron into selected ocean regions for the purpose of enhancing the oceans' uptake of carbon dioxide from the atmosphere (Martin et al. 1994). The big capacity of the oceans for sinking carbon dioxide from the atmosphere without any such intervention at the present time was described in Chapters 5 and 10 (Feely et al. 2020). The ocean iron fertilization may have undesirable consequences on the marine ecosystems, for which reasons the NRC experts assess the technology to be of too high risk, at the current stage of development.

Apropos of the Type III backstop technology, the cost of the nuclear fusion energy is assessed in the table. As I elaborated it in the previous section in terms of the Q efficiency, it is not yet effective as an alternative energy generation technology, notwithstanding the long-held high promises. At the present time, the nuclear fusion plant requires a larger amount of input energy than the amount of output energy it produces: $Q < 1$ (ITER 2020).

Having unveiled the costs of the backstop technologies, can the global community then proceed to deploying one of the Type I backstop technologies at a large scale as a global warming remedy? According to the experts at the NRC, the answer is no, which is highlighted in the last column of the table. The experts assess that it is feasible to put the technologies into practice but they further recommend that the technologies should not be tried at this time. The rationale for the recommendation is that a full risk assessment of such a large-scale experiment has not yet been performed satisfactorily while the side effects are expected to be substantial. Specifically, a creation of a layer of aerosols in the stratosphere of Earth may lead to unforeseen consequences on people and ecosystems.

To be more accurate from the economics standpoint, the risk is also an additional cost to the economy, that is, not apart from the total cost of the technology deployment. In other words, there is an additional cost element to the estimated costs of the various backstop technologies in Table 8.1, which may be incurred by an increased variability in the cost or an extreme cost event. The scientific/economics community should endeavor to put a dollar value on the risk of the backstops to help the policy decision while at the same time it works to minimize the risk.

To push one step further, a concern on the risk of a breakthrough technology should not hold back the global community from adopting and utilizing the technology when it is called for. The global community should strive to reduce the risk and, as long as the risk can be managed to a certain level, be prepared for deploying it if it is estimated that the benefit substantially outweighs the cost of the technology. This seems to be the case for the Type I backstop technologies in Table 8.1, which are estimated to cost only an order of the magnitude of the cost of the conventional decarbonization through regulations.

Having said that, it should be emphasized that the deployment of a backstop technology should be considered the last resort, that is, the backstop indeed when alternative mitigation options become far too costly for people to bear. If there are other policy options to go for whose costs are also modest, the global community should not need to rely on the backstops as the first resort.

8.5   ECONOMICS OF INCENTIVES: WHO WILL BUILD THE BACKSTOP TECHNOLOGIES?

The backstop technologies, if successfully developed and deployed, would provide an ultimate remedy for the Planet's climate problems. Even further, they would render humanity a control over the Planet's climate system, which will certainly be a landmark moment for the humankind. However, the cost of perfecting any of the breakthrough technologies in Table 8.1 is also enormous. Any inventor or country who would pursue one of these technologies should be prepared for spending billions of dollars without the full confidence that the promised technology will indeed be delivered within an agreed deadline. Consider the cost and promises of the nuclear fusion technology!

The question of who will develop such a costly backstop technology, or put differently, of how investments into such a costly technological innovation can be induced is at the heart of the economics of backstop technologies. This is where the term "greenhouse technologies" coined recently by the present author comes into play (Seo 2020).

To start with, notice that the farm greenhouse is built by a private investor who is seeking after economic returns from the investment. There is no other incentive needed, such as a government subsidy or a foreign development aid. Neither is there any need for a collaboration of a large number of investors to build a greenhouse. The investment into the technology arises purely from an economic incentive, that is, returns minus costs, for the investor.

Will one of the breakthrough technologies listed in Table 8.1 be developed purely by market forces alone, just like the way the farm greenhouses are built? Put differently, will an individual investor or inventor have an incentive to develop and perfect one of the backstop technologies purely from the economic viewpoint?

At first glance, such a prospect appears dim. Take for consideration the ongoing nuclear fusion project in Europe. The ITER project in France is a collaboration of 35 countries, whose cost estimates range from US$22 billion to 65 billion. The costly investment is beyond the reach of one individual investor or inventor. The project is led by the French government and the European Union.

However, once we start digging deeper into the question, a breakthrough technological innovation pertinent to the safeguarding the global climate system driven by an individual inventor does increasingly seem not unthinkable. Consider the Electric Vehicles (EVs) with an all solid-state-battery, the solar PV perovskite cells, and the Light Emitting Diode (LED) lamp. At the level of scientific inquiries and commercial interests that follow, there were, after all, ample incentives for the motivated scientists to develop and perfect these world-changing technologies (Akasaki et al. 2014; Reisch 2017; Tian et al. 2020). They are all considered a critical technology in the fight against the global warming risk for the Planet.

Of the backstop technologies enumerated in Table 8.1, let's consider this time the direct-air-carbon-capture technology. The technology has been researched and is being developed by individual researchers in an academic institution or another (Lackner et al. 2012). What motivates these researchers, in particular financially, to take the big risk involved?

The motivating forces are, inter alia, economic, research, and public interests. The economic incentive is the possibility of capturing carbon dioxide and selling it to, *e.g.*, agricultural producers who would use it for crop growth enhancements or builders who would make use of the captured carbon. The research incentive is that the technology is viewed as an important part of the biogeochemistry research. The public interest is the third incentive. Researchers are attracted by the overwhelming public interest in a climate control technology, including of the governments and the international organizations. The backstop technology is developed via investors' decisions to maximize the profit from the investment in the technology considering these incentives (Romer 1990).

Let's take a look at another backstop technology: a stratospheric aerosol layer (SAL). The science principle underlying the SAL technology became clear to scientists a long time ago. The Planet-cooling effect of aerosols in the stratosphere was identified in the earliest publications of the United Nations' reports (IPCC 1990; UNFCCC 1992). The process at the core of SAL is also naturally occurring, for example, the volcanic eruption that releases dust particles with the consequence of cooling the Planet. Scientists estimate that the volcanic eruptions of Mt. Pinatubo in the Philippines in 1991 had the cooling effect by about 0.2 degree Celsius.

The SAL technology is not only effective for cooling the Planet but also can be applied with necessary modifications to affect regional precipitations by enhancing regional cloud formations. The modified technology would be of great benefit to the local as well as national economies, beyond the benefit in terms of a global climate control option. Let me present two programs aiming at advancing such modified techniques for the economic benefits, which are summarized in Table 8.2, from the experiences of the China's weather modification offices. The first is the Rainmaker project and the second is the Sky River project.

First, shown as Type II in Table 8.2, the Chinese government is reported to be building a large number of mountain chambers across the Himalayan mountains for the purpose of climate engineering, especially for the purpose of increasing regional precipitations in the arid agricultural areas of northern China and adding water resources to the freshwater reserve of the Tibetan Plateau.

The "Sky River" project, as it is called by the Chinese, in the Tibetan Plateau is building tens of thousands of fuel-burning chambers over the Tibetan land areas as large as three times the size of Spain. The chambers will burn solid fuels to create silver iodide, a cloud seeding agent. Once emitted, the Indian monsoon winds from the South will carry the

Table 8.2 Weather modification technologies in China

Type	Technology specifications	Desired effects	Observed effects	Time
I	The Rainmaker: Seeding clouds by firing rockets and shells loaded with silver iodide into clouds	Make rain (1) to end droughts (2) to prevent hail storms (3) for firefighting (4) for countering the effect of severe dust storms	(1) Increased precipitation in Beijing by about 13% in 2004; (2) Nationwide, added 210 cubic kilometers of rain	(1) 2004 (2) 1995–2003
		Create snow	Created snow on New Year's Day	1997
		Make target places free of rain	Made the 2008 Summer Olympics free of rain	2008
II	The Sky River: (1) Tens of thousands of chambers built at selected locations across the Tibetan Plateau (2) The chambers burn solid fuel to produce silver iodide, a cloud-seeding agent (3) The monsoon wind from South Asia hits the mountains, sweeping the particles into the clouds to induce rain and snow	Increase the rainfall in the target region to supply by up to 10 billion cubic meters a year to the Tibetan Plateau, the Asia's biggest freshwater reserve		Circa 2018

particles to the clouds. The project's initial goal is to produce 10 billion cubic meters (m^3) of rainfall annually, most of which for the northern arid regions of China (SCMP 2018).

The second is the "Rainmaker" project, shown as Type I in Table 8.2. It was widely reported by the world media that the Chinese government created artificial rains before the Beijing Olympics in summer 2008 in order to remove air pollutants from the sky, thereby to provide clean air and clear sky during the period of the Olympic games. Further, it moved the rains to outside Beijing to make the Olympic games free of rain (Pontin 2008). A timeline of the different applications of the Rainmaker technique is summarized in Table 8.2.

Considering the beneficial outcomes observed or expected in the above-described weather modification programs, such technological capabilities are deemed undoubtedly to be of great value to local and national governments, setting aside the value to the global community, of China. In fact, the Chinese have long been fascinated for millenniums with the weather alteration capability as a crucial war strategy (Refer to the Record of Three Kingdoms). It is certainly possible that the governments of the US or of other countries decide sooner or later to invest substantially in a similar weather/climate modification technology for the economic or strategic reasons.

An individual inventor will not be permitted to alter the weather system of her country. However, as long as there is a strong governmental/pubic interest in the technologies similar to the Rainmaker and the Sky River, an individual researcher will advance and perfect them at a laboratory/field scale, which may be then scaled up when needed for the public purposes.

As long as the national-level weather control efforts would turn out to be successful and without harmful effects, which may be already the case as reported from China, Russia, and other countries, it may be only a matter of time to scale up the regional technological applications to the level which is meaningful for a global climate strategy.

The key take-away message from this section can be summarized as follows. There is an appropriate incentive for an individual inventor or investor to invest in the breakthrough backstop technologies for the global climate system, especially for the Type I and II backstop technologies in Table 8.1, considering a large range of economically-beneficial applications of the technologies. For the investor, it will be similar to inventing a sophisticated farm greenhouse for personal economic profits via enhancing plants' growth inside the structure, although the incentive mechanism for the backstops is more complicated owing to the public sector's role. At the fundamental level, a backstop technology investment is like a farm greenhouse investment. At the application stage of the technology through a public policy process, it is like a Sky River, which affects the society at large that adopts the technology.

8.6   YOUR LIFE AND FUTURE: THE END OF PLANETARY CIVILIZATIONS

The exposition of the greenhouse technologies in this chapter is given painstakingly by the present author. This extensive effort can be justified by the reality that the availability of backstop technologies is a critical variable for the assessment of the Planetwide risk that global warming realizations may pose in a not-far-away future. I am certain you have come to contacts quite often with the terms such as "climate emergency," "climate catastrophe," "unstoppable global warming," "only twelve years left" from the tipping point, or "the greatest threat to humanity" (US House of Representatives 2019). These expressions all point to the Planetary risk of climate change.

Of all such expressions, the direst one is perhaps "the end of all civilizations on the Planet as we know them" as the ultimate consequence of the unfolding global warming (Weitzman 2009). It is truly a doomsday possibility that the expression is getting us to. It is also a dark prediction on your life at present and the future that may hold for you. This is why we need to assess these declarative statements carefully and must not accept them blindly without examinations.

To assess the degree of the risk that global climate change could pose to the Planet, scientists rely most often on the probability that each possible climate change extreme event may occur. More precisely, we, like the scientists, need to be informed of the probability distribution across the range of possible climate change events. If the distribution could be made known, we would settle firstly on the mean (average) of all possible events as the best estimate of the risk. This estimate can then be used as an input for policy analyses and designs (Nordhaus 2008).

Having clarified that, the global community will be concerned about tail events besides the middle-of-the-range events, that is, the most extreme events in the range of possibilities. More specifically, many people would be interested in the probability that such an extreme climate event might occur. Are the tail probabilities too fat to be ignored? Or, are the tail probabilities too thin to be of any meaning? Economists coined the term a "fat-tail climate event" to refer to the tail-of-the-distribution event whose likelihood of realization remains significant however extreme the event is. The fat-tail event cannot be ignored and set aside from an analysis, to emphasize, however extreme the event is (Weitzman 2009).

Let me explain this more concretely to help your understanding of the main point of the fat-tail debate, which is not at all easy for anyone to comprehend at the first encounter. Let's pick a tail event: a 20° Celsius increase in global average temperature by the end of the twenty-first century from the twentieth century average temperature. First, it is certainly a tail event. As we repeatedly saw in this book, the best estimate of climate sensitivity is 3° Celsius according to the international group of scientists assembled by the Intergovernmental Panel on Climate Change (IPCC) (2013). Further, the range of possible climate change realizations suggested by the IPCC is from about 1.0 °C to 4.5 °C. Therefore, the 20 °C upward change in the global climate system through the next 100 years would be utterly shocking, let alone being unprecedented, and the eventual impact of the tail event shall amount to a Planetary collapse comparable to "the end of all civilizations on the Planet."

In the next step, we need to judge whether the tail event is a fat-tail event, a thin-tail event, or a medium-tail event. The defining distinction among these distributions is as follows. If it is a non-fat tail event, the probability of it being realized is approximately zero, that is, it would not occur in a statistical sense. If it is a fat-tail event, the probability of it coming to pass remains significant, meaning that it does not approximate zero in a statistical sense (Seo 2020).

Which of the two is the true tail distribution? This is where the family of backstop technologies enters the room like an elephant. Table 8.1 enumerated three types of the backstop technology and five distinct techniques in total. Of the five, the Type I technologies, i.e., the solar radiation management (SRM) technologies, are technically feasible now. At the same time, the total cost of deploying and operating the two techniques is not at all prohibitively large, in fact, far below the cost of the end of Planetary civilizations. Even with the full environmental risks accounted for, the total cost will most certainly be only a small fraction of the colossal damage comparable to the end of Earth.

Similarly, although the Type II technologies, referred to as the direct air capture and storage (DACS), are assessed by the experts to be somewhat costlier at the moment than the global mitigation policy comparable to the aspirations of the United Nations conventions, the total cost of the DACS deployment is once again not anywhere near the cost of a Planetary catastrophic end (Nordhaus 2008).

Considering the technical feasibilities and the total costs of the above-mentioned backstop technologies in this way, we can conclude with confidence that the probability of the concerned tail event, that is, a 20° Celsius increase in the global average temperature, is not significant. It may be concluded that the probability is near zero.

Let me state this crucial analytical result another way. The probability distribution of the range of possible climate change events can be estimated with and without the range of possible climate backstop technologies as well as the basket of microbehavioral incentives to invent and deploy them. The distribution estimated without them would look closer to a fat-tail distribution. On the other hand, the distribution estimated with them would look similar to a non-fat tail distribution whose tail probabilities decline in an "exponential" manner (Seo 2018, 2020).

8.7   Chapter Highlights

- This chapter explains the family of ultimate technologies that would render humanity a power of control, albeit partly, over the Planet's climate system, which are referred to as a greenhouse technology or a backstop technology.
- The history of a farm greenhouse shows that this simple innovation is the first climate control technology invented by humankind for the purpose of gaining economic profits and has become ever more powerful through the centuries.
- There are three types of the backstop technology: solar radiation management, a direct air capture and storage, and a zero-carbon energy generation.
- The economic and policy questions on the backstop technologies are centered around their costs of deployment and operation as well as the risks of a Planetwide experiment.
- An even more critical economic question is whether there is an incentive for an individual inventor/investor to develop one of these ultimate technologies. In this regard, this chapter elucidates various weather/climate control programs from China such as the Rainmaker and the Sky River.

- Taking into consideration of the technical feasibilities and the total costs involved in the backstop technologies, a tail event such as a 20 °C increase in the Planet's temperature cannot be assessed to be of a fat tail. Stated differently, the possibility of the end of all Planetary civilizations caused by climate change would approximate zero.

References

Akasaki, Isamu, Hiroshi Amano, and Shuji Nakamura. 2014. Blue LEDs: Filling the World with New Light. Nobel Prize Lecture. Stockholm, SE: The Nobel Foundation. Accessed form http://www.nobelprize.org/nobel_prizes/physics/laureates/2014/popular-physicsprize2014.pdf.

Feely, R.A., R. Wanninkhof, P. Landschützer, B.R. Carter, and J.A. Triñanes. 2020. Global Ocean Carbon Cycle [in State of the Climate in 2019]. *Bulletin of the American Meteorological Society* 101 (8): S170–75.

Greenhouse Lighting & Systems Engineering (GLASE). 2019. *Growing the World's Food in Greenhouses*. Ithaca, NY: The GLASE, Cornell University.

Intergovernmental Panel on Climate Change (IPCC). 1990. *Climate Change: The IPCC Scientific Assessment*. Cambridge, UK: Cambridge University Press.

Intergovernmental Panel on Climate Change (IPCC). 2013. *Climate Change 2013: The Physical Science Basis*. The Fifth Assessment Report of the IPCC. Cambridge: Cambridge University Press.

International Thermonuclear Experimental Reactor (ITER). 2015. *ITER: The World's Largest Tokamak*. Paris, FR: The ITER. Accessed form https://www.iter.org/mach.

International Thermonuclear Experimental Reactor (ITER). 2020. *ITER: The World's Largest Tokamak*. Saint-Paul-lès-Durance, Fr.

Lackner, K.S., S. Brennana, J.M. Matter, A.A. Park, A. Wright, and B.V. Zwaan. 2012. The Urgency of the Development of CO2 Capture from Ambient Air. *Proceedings of the National Academy of Sciences, USA* 109 (33): 13156–62.

Lawrence Livermore National Laboratory (LLNL). 2015. *How NIF Works*. Berkeley, CA: The LLNL. Accessed form https://lasers.llnl.gov/about/how-nif-works.

Le Treut, H., R. Somerville, U. Cubasch, Y. Ding, C. Mauritzen, A. Mokssit, T. Peterson, and M. Prather. 2007. Historical Overview of Climate Change. In *Climate Change 2007: The Physical Science Basis*, ed. S. Solomon et al. The Fourth Assessment Report of the Intergovernmental Panel on Climate Change. Cambridge: Cambridge University Press.

Martin, J.H., K.H. Coale, K.S. Johnson, S.E. Fitzwater, et al. 1994. Testing the Iron Hypothesis in Ecosystems of the Equatorial Pacific Ocean. *Nature* 371: 123–29.

Muijzenberg, Erwin W.B. van den. 1980. *A History of Greenhouses*. Wageningen, NL: Institute for Agricultural Engineering.

National Renewable Energy Laboratory (NREL). 2020. *Perovskite Solar Cells*. Washington, DC: The NREL, Department of Energy. Accessed from https://www.nrel.gov/pv/perovskite-solar-cells.html.

National Research Council (NRC). 2015a. *Climate Intervention: Reflecting Sunlight to Cool Earth. Committee on Geoengineering Climate: Technical Evaluation and Discussion of Impacts*. Washington, DC: The National Academies Press.

National Research Council (NRC). 2015b. *Climate Intervention: Carbon Dioxide Removal and Reliable Sequestration*. Washington, DC: The National Academies Press.

Nordhaus, William. 1973. The Allocation of Energy Resources. *Brookings Papers on Economic Activities*: 529–76.

Nordhaus, William D. 2008. *A Question of Balance—Weighing the Options on Global Warming Policies*. New Haven, CT: Yale University Press.

Nordhaus, W. 2013. *The Climate Casino: Risk, Uncertainty, and Economics for a Warming World*. New Haven, CT: Yale University Press.

Nuclear Energy Institute (NEI). 2016. *Energy Statistics*. Washington, DC: The NEI. Accessed from http://www.nei.org/Knowledge-Center/Nuclear-Statistics.

Pontin, M.W. 2008. *Weather Engineering in China*. Cambridge, MA: MIT Technology Review. Accessed from https://www.technologyreview.com/s/409794/weather-engineering-in-china/.

Reisch, Mark S. 2017. Solid-state Batteries Inch Their Way toward Commercialization. *Chemical and Engineering News, American Chemical Society* 95 (46): 19–21.

Romer, Paul M. 1990. Endogenous Technical Change. *Journal of Political Economy* 98: S71–S102.

Seo, S. Niggol. 2018. *Natural and Man-made Catastrophes: Theories, Economics, and Policy Designs*. Hoboken, NJ: Wiley-Blackwell.

Seo, S. Niggol. 2020. Appendix: A Succinct Mathematical Disproof of the Dismal Theorem of Economics. In *The Economics of Globally Shared and Public Goods*. Amsterdam, NL: Academic Press.

South China Morning Post (SCMP). 2018. *China Needs More Water. So It's Building a Rain-making Network Three Times the Size of Spain*. Hong Kong: SCMP. Published on March 26, 2018.

Tian, Xueyu, Samuel D. Stranks, and Fengqi You. 2020. Life Cycle Energy Use and Environmental Implications of High-performance Perovskite Tandem

Solar Cells. *Science Advances* 6 (31): eabb0055. https://doi.org/10.1126/sciadv.abb0055.

Tol, R. 2009. The Economic Effects of Climate Change. *Journal of Economic Perspectives* 23, 29–51.

United Nations Framework Convention on Climate Change (UNFCCC). 1992. *United Nations Framework Convention on Climate Change.* New York: UNFCCC.

Unites States House of Representatives. 2019. *Resolution: Recognizing the Duty of the Federal Government to Create a Green New Deal.* Washington, DC: United States House of Representatives. Published February 7, 2019.

Viviano, Frank. 2017. *How the Netherlands Feeds the World.* Washington, DC: National Geographic, September.

Weitzman, M. L. 2009. On Modeling and Interpreting the Economics of Catastrophic Climate Change. *Review of Economics and Statistics* 91 (1): 1–19.

Wittwer, S.H., and W.M. Robb. 1964. Carbon Dioxide Enrichment of Greenhouse Atmospheres for Food Crop Production. *Economic Botany* 18 (1): 34–56.

Yoon, Sang J., and Jan Woudstra. 2007. Advanced Horticultural Techniques in Korea: The Earliest Documented Greenhouses. *Garden History* 35: 68–84.

CHAPTER 9

A Story of Polar Bears and Penguins: A Paradox of Biodiversity and Climate Change

9.1 Introduction

I had the chance to visit the city of Winnipeg in the province of Manitoba in Canada, which is situated in the central landlocked heartland of the country. It felt extremely cold even to me who had 'adapted' to a cold winter of New England well. I was told Winnipeg is the coldest city in Canada during the wintertime and later learned that the city ranks first in the number of days with daily average temperature below minus 30 °C. Having considered the vast frozen lands in the country's North, say, permafrost zones of the country, I asked a group of locals whether they would be happy with an ever-warmer Earth in the coming decades. To my surprise, they replied with no, expressing that they are very concerned about polar bears in the north of the country!

An image of a polar bear on a small piece of ice drifting in the Arctic Ocean, with a loss of her long-held habitats because of a melting of the Arctic ice caps and sheets, is arguably the most iconic image in the pantheon of global warming communications. The image may appear with a caption warning that polar bears will lose their habitats completely if the hotter Planet should make the Arctic Ocean ice-free.

In Fig. 9.1, a dramatic change in the sea ice extent of the Arctic Ocean is illustrated. It shows the summer ice minimum of 1979 as well as the summer ice minimum of 2019. The overlapping map shows a large decrease in the ice extent of the Arctic. The data are made available by

© The Author(s), under exclusive license to Springer Nature Switzerland AG 2021
S. N. Seo, *Climate Change and Economics*,
https://doi.org/10.1007/978-3-030-66680-4_9

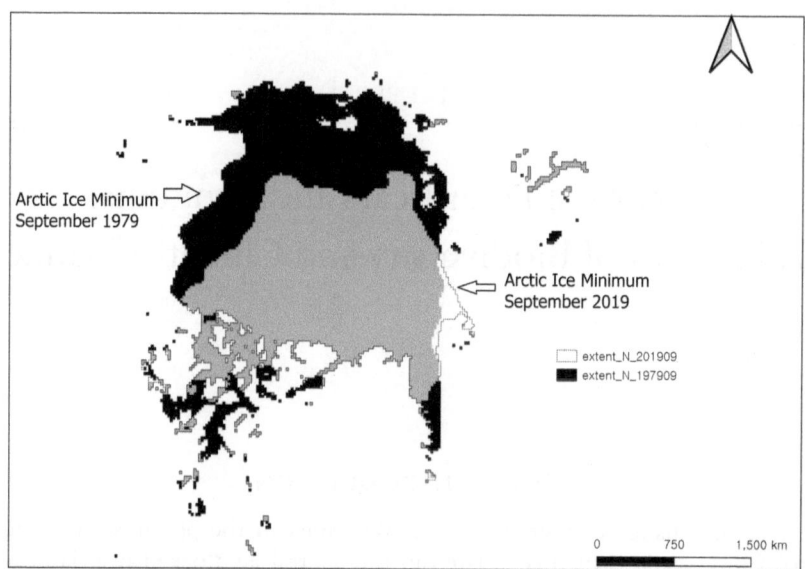

Fig. 9.1 A dramatic decrease in the Arctic ice extent

the National Snow and Ice Data Center (NSIDC) of the US which relies on, for example, the Special Sensor Microwave Imager/Sounder (SSMIS) on board the Defense Meteorological Satellite Program (DMSP) satellites (Fetterer et al. 2017). The white areas are the ice sheets in 2019 while the black-shaded areas are the ice sheets in 1979. The gray shade is the overlap of the ice sheets of the two years.

To a young climate activist like yourself, a forecast of an eternal disappearance of a certain species of animals, birds, fish, and even insects from the Planet caused by human-induced global warming and climate change would be an emotionally distressing destiny for you to accept (*TIME* 2020). An even more depressing question to ponder is whether humanity should feel guilty and be blamed if global warming were to make the presently cold-climate countries warmer and more pleasant for humans to live but, at the same time, were to drive the polar bears starved to death.

The science of climate change is often quite dismal, rooting on or for the frightening terms such as 'the end of Planetary civilizations,' 'Planetary emergency,' 'a runaway catastrophe,' and 'a tipping point of no return.' These proclamations are depressing to any citizen of the Planet, but especially to the young people. One of such dismal predictions is a mass extinction of species. A mass extinction has occurred multiple times in the geologic time scale of the Planet (Meyer and Kump 2008). A dismal scientist is often quite confident that a sixth mass extinction will occur ineluctably through a global warming and climate disruption (Ceballos et al. 2017).

Should we be worried about such a dismal event? How much should we be worried? Before we can make any definitive decision, a rigorous review of the scientific literature must be conducted first. It may or may not turn out to be that such a dismal prediction of a mass extinction is supported by the balance of all evidence and modeling efforts (Mayhew et al. 2008; Blowes et al. 2019). We should certainly hope that the forecast of a mass extinction be at fault or at odds with the reality and the scientific knowledge, which would be good news and a relief to the citizens of the Planet.

Only after being based on a solid review of the science and being confident on the scientific models, we will be able to proceed to tackle a host of critical policy questions, including the following: What can be done by the global community or an individual nation to preserve Planetary biodiversity richness?; What specific steps are needed to protect an individual species, such as polar bears, against extinction? These questions are at the heart of the international efforts coordinated by the Convention on Biological Diversity (CBD) to which nearly 200 nations are signatories (CBD 1992; IUCN 2020).

At the core of these questions in turn lie some of the thorniest intellectual problems pertaining to the economics and policy of conservation and preservation of species (Simpson et al. 1996; Weitzman 1998). To mention some, would the world citizens choose to protect all species against extinction, despite they are informed that a large number of new species will simultaneously originate without such preservation efforts? If the world community should choose to prioritize in its conservation decision, considering the economic reality of high costs of conservation programs, which species should be chosen for protection and which not? What factors must be considered and how much weights should be given to them in making such choices? What measures and strategies would achieve the best outcome in protecting a certain species?

9.2   Biodiversity and the Treaty

Biodiversity is an encompassing concept as will be clarified presently, which makes it a central indicator in many academic inquiries such as species richness, endangered species, a mass extinction, conservation of species, and a loss of habitat of polar bears and other species. Biodiversity, which is a short form of biological diversity, is defined by the Convention on Biological Diversity (CBD) of the United Nations as "the variability among living organisms" and "the ecological complexes that they are part of." The living organisms from all sources are taken into account including, *inter alia*, terrestrial, marine, and other aquatic ecosystems. The concept of biological diversity embraces diversity within species, diversity between species, and diversity of ecosystems (CBD 1992).

The Convention on Biological Diversity (CBD) is the first global treaty, signed by 196 parties as of 2020, whose primary mission is "conservation of biological diversity." The CBD, often referred to as the Biodiversity Convention, is also an international organization under the United Nations. The CBD was opened for signature at the Rio Earth Summit in 1992, like the United Nations Framework Convention on Climate Change (UNFCCC), and entered into force the next year. The declared objectives of the Convention in 1992 are the conservation of biological diversity, the sustainable use of its components, and the fair and equitable sharing of the benefits arising out of the utilization of genetic resources. Notably, the CBD is the first treaty that stated that the conservation of biodiversity is a common concern of humankind (CBD 1992).

Of the most recent developments regarding the Convention, the first draft of the post-2020 framework of the Biodiversity Convention, which was prepared for a global summit scheduled later in September 2020 in New York, calls for 30% of all land and sea areas of the Planet to be protected by 2030 through designations of protected areas across the Planet (CBD 2020). The draft notes that the CBD is alarmed by the continued loss of biodiversity and the threat it poses to human well-being.

The CBD's goals are set with reference to the goals in the UN climate change conferences, specifically, the temperature target declared in the Paris Agreement to avoid a dangerous global climate change. To meet the aspirational target of the Paris Agreement of keeping global temperature increase by 1.5 °C, some scientists even propose to protect 50% of all land and sea areas on the Planet (UNFCCC 2015; Dinerstein et al. 2019).

To any citizen on the Planet, I feel that the proposal of designating half the Planet as protected areas for the sake of biodiversity conservation would be felt as an extremely ambitious goal as well as a gigantic sacrifice by humankind, even after considering the vast ocean areas which are uninhabited by people. Should the human community accept and go with the proposal? Are we planning to sacrifice too much for the sake of species richness?

These are intriguing yet important questions. To arrive at a satisfactory answer to these questions, we will have to start with reviewing the up-to-date scientific and economic evidence, which we will do in the next sections.

9.3   THE PARADOX OF BIODIVERSITY AND GLOBAL WARMING

If you happen to have visited the Amazon rainforest in South America, it may have dawned on you that there is something odd about the grave concerns about the biodiversity loss and a mass extinction of species caused by a globally warmer Planet, such as those inscribed in the Biodiversity Convention. In your trip to the Amazon, you would have encountered a greater diversity of life forms as well as a greater ecological complexity there than anywhere else on the Planet, which would most certainly have amazed you bigly.

Of the places on Earth that are famed for their great diversity of species of plants and animals are the Amazon rainforest in Latin America, the Rwenzori National Park at the edge of the Congo River in Central Africa, the Javanese rainforest in Indonesia, and the Island of Ceylon (also called Sri Lanka) in the Indian Ocean. The commonality across these celebrated places of the richest biodiversity is a hot climate system around the Equator.

If the richest biodiversity spots are located in the hot climate zones, how is it possible that the warmer Planet in the future will lead to a large decrease in biodiversity or even a mass extinction of species? If it were a puzzling observation to you, you are not alone. This has been described as the paradox of biodiversity among the climate ecologists (Mayhew et al. 2012).

Researchers who mapped the richness of biodiversity across the Planet, that is to say, across the latitudinal zones on Earth, find that the lower the latitude of a region is, the richer the biodiversity of it becomes; alternatively, the higher the latitude of a region is, the poorer the biodiversity of it becomes.

To give you a visual example, let's imagine the number of species is mapped across the American continent. In the Northern Hemisphere: a region from Alaska to Canada, from Canada to the US, from the US to Mexico, from Mexico to Costa Rica and Panama. In the Southern Hemisphere: a region from the southern tip of Argentina to the Pampas grasslands and Uruguay, from Uruguay to the Brazilian Cerrado, from the Brazilian Cerrado to the Amazon rainforest and the Andean highlands. Many biologists have mapped a decreasing biodiversity richness from lower-latitude regions to higher latitude regions in the American continent as well as across the Planet. This is formally referred to as the latitudinal biodiversity gradient (LBG). According to Willig and his colleagues, the LBG is "the oldest and one of the most fundamental patterns concerning life on earth" (Willig et al. 2003).

Notwithstanding, this is not the end of the story on the biodiversity gradient. For concerned scientists, an alternative methodology for establishing the association between a global temperature increase and a change in biodiversity richness is to examine the changes in biodiversity richness across the geologic times of the Planet. The Earth's temperature has undergone cyclical and abrupt changes across the Paleozoic, Mesozoic, and Cenozoic eras, all of which belong to the Phanerozoic Eon. To any sentient being, this is a very long lapse of time, stretching half a billion years. Climate scientists devised a methodology for measuring the biodiversity richness at one tiny point in the ceaseless geologic timescale, even from many hundred million years ago, as well as the changes of it across the timescale, relying on the fossils formed across these geologic times.

Applying this methodology, which is a time-series analysis, to a large pool of fossil records that covers 540 million years in the Earth's history, a group of scientists reported a gradual decrease in the biodiversity richness when the Planet swings toward a warmer climate period. In addition, they reported that the extinction rate of species becomes higher as the Planet swings from a colder climate period to a warmer climate period, which occurred many times across the geologic times (Mayhew et al. 2008).

What a surprise! How does it come to be that the latitudinal studies and the time-series studies are arriving at the completely opposite conclusions on the same question? This was referred to as the paradox of biodiversity and climate change, which has puzzled not only ecologists but also other observers of climate science and activists: Why warm temperatures should decrease biological diversity through time (across the geologic timescale) while warm temperatures should increase biological diversity across space (across the latitudinal gradient of Earth). Can the paradox be explained?

9.4 Will Global Warming Reduce Biodiversity Richness?

Has a global warmer period in the past reduced biodiversity richness? Will the future Planet warming decrease the biodiversity richness and increase the extinction rate? Acknowledging the paradox above explained, you and I cannot make any reasonable conclusion on these critical questions for policy designs, let alone being bullish and forceful as an activist. Then, there came a breakthrough: selection bias.

The problem of selection bias is pervasive in the scientific and economic models of global warming, although the theory was first originated in the economics of labor markets four decades ago (Heckman 1979). The concept was elucidated markedly through the economics of climate change in the past two decades (Seo 2006, 2016). In the context of biodiversity under consideration in this chapter, the selection bias means that the fossil records of the geologic times that were examined by the scientists were selected from some parts of the Planet while many other parts of the Planet were set aside, *albeit* unintentionally. A selective sampling such as this of the fossil records for an analysis is certain to bias the final outcomes of the analysis, including estimations and predictions. More subtly, a selective sampling is in most cases occurring unintentionally, that is, not by an intentional maneuver of the scientists but by their ignorance about the underlying processes.

To be concrete, the selection problem in the above-referenced time-series study was created owing to, on the one hand, the differing profiles of sedimentary rocks in the samples of different geologic times and, on the other hand, ignorance about the underlying process in which the amount of sedimentary rocks itself determines the paleo-biodiversity richness. Without a correction of the sampling bias by the researchers of the above, a high biodiversity richness in the sample would simply reflect a larger amount of sedimentary rocks while a low biodiversity richness in the sample would just reflect a smaller amount of sedimentary rocks in the fossils. The scientific predictions are, therefore, as long as the selection problem is pervading in the sample, of no precision when it comes to unveiling a true climate-biodiversity relationship.

This pivotal shortcoming finally dawned on concerned scientists. After controlling the selectivity bias and focusing on the marine invertebrate biodiversity patterns across the Phanerozoic Eon (covering Paleozoic, Mesozoic, and Cenozoic), they were able to arrive at a solid conclusion, thereby, reconciling the predictions from the time-series studies with those from the latitudinal studies. They find that a warmer temperature period is associated with a higher biodiversity richness, reversing the previous predictions (Mayhew et al. 2012). Further, the researchers reported that a warmer temperature period is associated with a higher extinction rate, but at the same time a higher origination rate of species.

At this point, you might be inclined to ask whether such a reliable conclusion can be obtained only from the fossil records which are as old as several hundreds of millions of years. This is a reasonable reservation. Attending to the concern such as yours on the geologic timescale data, an ecologist may approach the biodiversity and climate question with a novel alternative methodology. To be specific, the researcher can rely on the recent biodiversity data, for example, a 40-year time-series data of a biodiversity index or another. In addition, she/he can rely on the time-series data obtained from a specific geographical zone, which can be repeated across a large number of geographical zones of the Planet.

In this spirit, a group of biologists recently analyzed the 239 independent studies that examined the changes in the biodiversity indices at a specific geographic zone or location, based on the data compiled by the BioTIME database (Dornelas et al. 2018; Blowes et al. 2019). Some data in the BioTIME database go back to the 1800s, but most data cover the time period of the past 40 years. In total, the researchers analyzed more than 50,000 time series of a biodiversity index or another.

The group concludes that assemblage richness, which is defined to be species richness at the local level, is not changing on average over past 40 years, although the Planet has gradually warmed over this time period. In more detail, the average assemblage richness does not change, although some regions are experiencing an increase in assemblage richness while other regions are experiencing a decrease in assemblage richness. The result holds for marine, terrestrial, and freshwater realms, respectively.

The group also confirms other changes pertinent to biodiversity richness. The above conclusion holds despite a rapid compositional change in species assemblage which is more evident in marine biomes than in terrestrial biomes. In addition, the authors report a rapid increase or decrease in assemblage richness by as much as 20% per year in an individual time series of assemblage richness (Blowes et al. 2019).

The conclusions from this analysis of the location-specific and contemporary time-series data appear to concur largely with the conclusions from the analysis of the geologic timescale fossil data with a proper selectivity bias correction (Blowes et al. 2019; Mayhew et al. 2012). The agreement of the results derived from the two different scientific methodologies as well as the latitudinal biodiversity gradient, gives us higher confidence on a true climate-biodiversity relationship, even if we could not be 100% sure at this point in time and further examinations are warranted.

9.5 Should We Save Polar Bears and Penguins?

The review so far does not paint a bleak picture on the future of biological diversity under a warmer Planet. On the contrary, the review indicates that global warming may increase the species richness of animals, fish, and plants on the Planet. The question, however, is still unanswered on the well-being of polar bears which was our starting point of this chapter. Even if the global biodiversity richness should hold up in a warmer Planet, a certain species, such as polar bears, may be unable to avoid extinction. In the case of such species, should humanity strive to save the individual species at risk of extinction or let the nature take care of the process of species turnovers in a changing climate?

To clarify this question, let's begin with the situations faced by polar bears and penguins. The survival and risk faced by these well-loved animals depend on how and how fast the ice sheets in the polar regions—the Arctic, the Antarctic, and the Greenland—will melt in the coming decades because of a gradually warming Planet. An extinction of polar bears would be possible if the ice sheets in all three polar regions were to melt completely during the summer season.

If, on the other hand, only one of the three polar regions were to become free of ice during the summer season, polar bears and penguins in the other two regions with remaining ice sheets would be able to survive the summer. Polar bears in the ice-free Arctic Ocean could be relocated to the other polar regions, say, the Greenland, still with remaining ice sheets. From another angle, the entire ice sheets in the Arctic Ocean need not be ice-covered fully in order for polar bears to survive there.

Pursuing this idea further, a key statistic for the assessment of the survival risk faced by polar bears is the Arctic sea ice minimum. The Arctic sea ice, a primary habitat for polar bears which was depicted in Fig. 9.1, expands during the winter and shrinks during the summer. The Arctic sea ice reaches its minimum in every September. The National Aeronautics and Space Administration (NASA) database shows that the September sea ice extent in the Arctic is declining at a rate of 12.85% per decade, from the baseline of the 1981 to 2010 average ice extent (NASA 2020).

Figure 9.2 shows the changes in the Arctic sea ice extent in every September from 1979 to 2019 with an overlay of a linear trend line. The data are again from the National Snow and Ice Data Center of the US government (NSIDC 2020). The size of the ice extent has declined from roughly 7 million square kilometers during the first decade of the record, that is, the 1980s, to roughly 5 million square kilometers in the last decade of the record, that is, the 2010s.

What has happened to the Antarctic ice sheets or the Greenland ice sheets? According to the NASA, the Antarctica, a primary habitat for

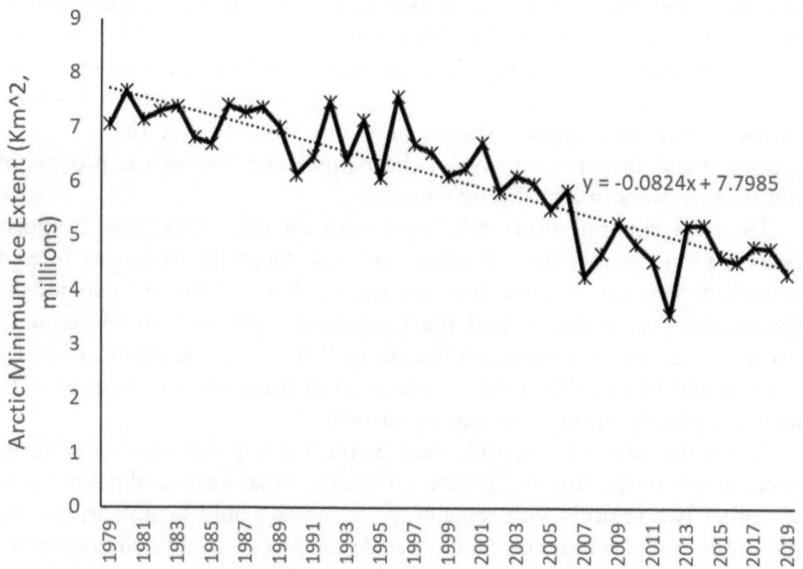

Fig. 9.2 Arctic sea ice minimum

penguins, has lost 2,500 gigatons of ice mass during the 2002–2020 period, annually 146 gigatons of ice mass lost (NASA 2020). From the 1979 to 2002 period, the rate of decrease was far slower, concretely, about 50 gigatons ice mass lost per year. This may seem like a massive loss to you. In terms of the total mass, however, which is about 26.5 million gigatons of ice mass in the Antarctica, the total ice mass loss during the first two decades of twenty-first century amounts to only 1/10,000 of the total mass, that is, one hundredth of 1%.

More intriguing is that, on the opposite pole of the Planet, the Antarctic ice extent has unveiled a trend, shown in Fig. 9.3, at odds with that of the Arctic September minimum ice extent in Fig. 9.2. The Antarctic September ice extent has revealed an increasing trend in the Antarctica from 1979 to 2014, which puzzled climate scientists (NSIDC 2014). For the entire period from 1979 to 2019, the time series of the September sea ice extent for the Antarctica shows a positive trend: +1.3% per decade (NSIDC 2020).

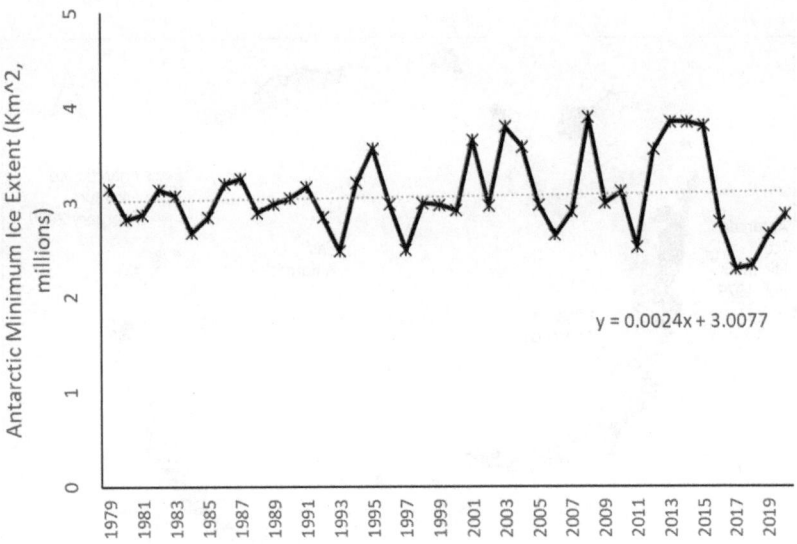

Fig. 9.3 Antarctic ice minimum extent (February)

In Fig. 9.3, you can verify the year-to-year changes in the Antarctic minimum ice extent for the 1979 to 2020 period. In Southern Hemisphere, a summer ice minimum is reached in every February of each year. The figure shows that the summer ice minimum extent in the Antarctica is about 3 million square kilometers. A linear trend is overlaid to the yearly changes, which shows a positive linear trend, despite the large drops observed in the summers of 2016 and 2017.

The figure conveys that the Antarctic ice sheets have been quite stable despite the observed global warming trend during the past 41-year time period captured in the figure. This stability may be attributed to several factors, including a relatively slower rate of warming in the Southern Hemisphere. A further examination of the Antarctica reveals that there are two sub-regions which have exhibited dissimilar trends in the ice extent: West Antarctica and East Antarctica, which is another contributing factor to the trend.

As the spatial map of the Antarctica in Fig. 9.4, drawn from the aforementioned NSIDC data set, reveals (Fetterer et al. 2017), the West

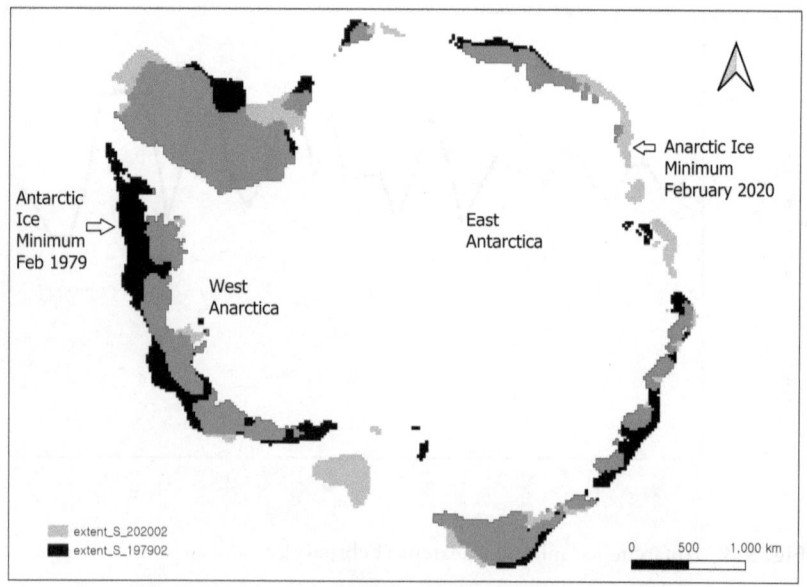

Fig. 9.4 Changes in the Antarctic ice extent: West and East Antarctica

Antarctic ice sheets, which have attracted keen interests from climate scientists, have shrunk through this time period (Oppehheimer 1998). The black shade is the February ice extent in 1979 and the gray shade is that in 2020. The dark gray shade is the overlap of the two-year ice sheets. By contrasts, the East Antarctic ice sheets have increased substantially over the 41-year time period according to the record measured by the aforementioned Defense Satellite of the US for the NSIDC. The Special Microwave Sensors and Radiometers aboard the satellite are capable of tracking the changes in the sea ice extent. To be specific, the gray shaded areas in the East Antarctica in the figure are newly expanded ice sheets during this period.

Other than the three ice-covered polar regions, polar bears can also survive in the land ecosystems, for example, permafrost zones in Canada and Alaska in North America, Siberia in Asia, and Scandinavia in Northern Europe. As such, even in the worst-case scenario of a complete melting of the Arctic Ocean during the summertime, it could be imagined that polar bears be relocated to the ice-covered above-mentioned land ecosystems, with an appropriate governmental protection given during the summer season. In fact, polar bears are sometimes found in these permafrost regions and brown bears, the closest relative to polar bears, live in these land ecosystems.

To sum up, the ensemble of evidence and predictions explained thus far points to us that the possibility of polar bears or penguins becoming extinct in this century owing to a Planet-wide warming is quite slim, taking for granted that the current trends in the various indicators introduced in Figs. 9.2, 9.3, and 9.4 would continue to hold.

Having said that, can we draw the same conclusion with regard to all other vulnerable species of animals? The answer may be no for some species. We can draw this conclusion because multiple scientific studies predict a higher rate of total turnover in species richness in a warmer world: that is, simultaneously a higher rate of extinction and a higher rate of origination (Blowes et al. 2019). These studies inform us that there will be a basket of species that will disappear owing to the Planet warming, although there will be another (larger) basket of species that will originate owing to the Planet warming.

Taking into consideration both the higher rate of total turnover and the expected change in biodiversity richness under a warmer Planet, should the human community intervene to save an endangered species that is predicted to be extinct without public interventions, for example, monarch butterflies? If it should, which of the basket of the threatened species should be the first to be chosen for conservation? If it should, how much financial resources should it set aside spending on conservation programs (Weitzman 1998)?

This is an onerous question to handle and on which public opinions are varied widely. An answer in one end of the spectrum of opinions is to let the nature take care of the course of species richness: that is, let it be and untouched as long as there is an overall increase in the biodiversity richness caused by global warming. At the other end of the spectrum of opinions, however, it will be argued on an ethics basis that each and every threatened species must be protected against extinction especially whose primary cause is human activities.

The two extreme positions will turn out, I predict, to be equally difficult for one to defend for one reason or another. A middle ground position between the two extremes will be sought and in fact we may be already most familiar with this middle ground position. Consider this middle ground position: Let it be as long as there is an overall increase in biodiversity richness, but, at the same time, let us protect a basket of species selectively against becoming extinct. The set of criteria for the selection of a species to be protected, the most important task if we would take this alternative position, may include, inter alia, people's desire to save the species, genetic importance of the species in the ecology of the Planet, and medical values (Simpson et al. 1996).

You may feel that the middle ground position stands, at least at first sight, a better chance of being agreed upon by people of divergent opinions. As the next Chapter will unravel, however, the middle ground position cannot escape spawning a new set of grave challenges to policy designers and theorists. The present author will come back in Chapter 10 to clarify these challenges and explain how they may be addressed conceptually and in a policy setting.

9.6   Final Words

An extinction of a species has long been a volatile topic, expressed lucidly by such classic writings as Aldo Leopold's "A Sand County Almanac" and Rachel Carson's "Silent Spring" (Leopold 1949; Carson 1962). The former tells it through "a fierce green fire in the eyes of a dying wolf" and the latter through "a spring without singing birds." A mass extinction of species has happened multiple times in the past of the Planet's history.

A prediction of another mass extinction caused by the human-induced Planet warming, which is often copiously featured in the media and political discourses, has stirred up volatile emotions among the people, especially the young.

I would like to, and am in a position to, end this chapter with a positive note. The review and analyses presented in this chapter point us to a starkly different conclusion and outlook with regard to the climate change-induced biodiversity losses and mass extinction. As far as the subject thereof is concerned, rational minds should triumph emotional responses. To the rational minds, scientific advances, some of which are elaborated in this chapter, will have a lot to offer in planning appropriate actions, policy or individual. Although we should be caring about each and every species of animals, we are simultaneously forced to accept an ever-evolving nature of the biosphere.

9.7   CHAPTER HIGHLIGHTS

- Will a mass extinction of species occur owing to a human-induced global warming? Should we be concerned about the extinction of polar bears and other vulnerable species? These are the questions that have long captivated the public.
- The international efforts to preserve biodiversity can be traced back to the Convention on Biological Diversity (CBD) of the United Nations, which declares the biodiversity preservation as the common concern of humankind.
- Despite the grave concern on biodiversity and mass extinction, scientists have long pondered over the paradox of biodiversity richness and global warming: Across the Planet, the hotter a region, the richer the biodiversity.
- Recent scientific advances reconcile the latitudinal studies and the time-series studies by addressing the sample selection problem in fossil records. They suggest that a warmer Planet may not decrease biodiversity richness, but an extinction of some species may be unavoidable.
- The satellite records of the polar ice sheets over the past four decades reveal that there has been an increase in the Antarctic sea ice extent and a decrease in the Arctic sea ice extent. If the Arctic Ocean and the Greenland were to become free of ice during the summertime,

polar bears would face extinction owing to a complete loss of their habitats.

- An onerous policy question is which of the endangered species should be first preserved while others be let go of to extinction as well as whether the human community should protect each and every threatened species from extinction.

REFERENCES

Blowes, Shane A., Sarah R. Supp, Laura H. Antão, Amanda Bates, Helge Bruelheide, Jonathan M. Chase, Faye Moyes, et al. 2019. The Geography of Biodiversity Change in Marine and Terrestrial Assemblages. *Science* 366 (6463): 339–45.

Carson, Rachel. 1962. *Silent Spring*. Boston, MA: Houghton Mifflin.

Ceballos, Gerardo, Paul R. Ehrlich, and Rodolfo Dirzo. 2017. Biological Annihilation via the Ongoing Sixth Mass Extinction Signaled by Vertebrate Population Losses and Declines. *Proceedings of the National Academy of Sciences* 114 (30): E6089–96.

Convention on Biological Diversity (CBD). 1992. *Convention on Biological Diversity*. New York, NY: United Nations.

Convention on Biological Diversity (CBD). 2020. *Zero Draft of the Post-2020 Global Biodiversity Framework*. New York, NY: United Nations.

Dinerstein, E., C. Vynne, E. Sala, A.R. Joshi, S. Fernando, T.E. Lovejoy, J. Mayorga, et al. 2019. A Global Deal For Nature: Guiding Principles, Milestones, and Targets. *Science Advances* 5 (4): eaaw2869. https://doi.org/10.1126/sciadv.aaw2869.

Dornelas, Maria, Laura H. Antão, Faye Moyes, Amanda E. Bates, Anne E. Magurran, Dušan Adam, Asem A. Akhmetzhanova, et al. 2018. BioTIME: A Database of Biodiversity Time Series for the Anthropocene. *Global Ecology and Biogeography* 27 (7): 760–86. https://doi.org/10.1111/geb.12729.

Fetterer, F., K. Knowles, W.N. Meier, M. Savoie, and A.K. Windnagel. 2017. *Sea Ice Index, Version 3*. Boulder, CO: National Snow and Ice Data Center. https://doi.org/10.7265/N5K072F8. Accessed on June 23, 2020.

Heckman, J. 1979. Sample Selection Bias as a Specification Error. *Econometrica* 47, 153–62.

International Union for the Conservation of Nature (IUCN). 2020. *The IUCN Red List of Threatened Species*. Version 2020-2. Gland, CH: IUCN. https://www.iucnredlist.org.

Leopold, Aldo. 1949. *A Sand County Almanac: And Sketches Here and There*. Oxford: Oxford University Press.

Mayhew, Peter J., Gareth B. Jenkins, and Timothy G. Benton. 2008. A Long-Term Association Between Global Temperature and Biodiversity, Origination and Extinction in the Fossil Record. *Proceedings of the Royal Society B: Biological Sciences* 275 (1630): 47–53. https://doi.org/10.1098/rspb.2007.1302.

Mayhew, Peter J., Mark A. Bell, Timothy G. Benton, and Alistair J. McGowan. 2012. Biodiversity Tracks Temperature over Time. *Proceedings of the National Academy of Sciences* 109 (38): 15141–45. https://doi.org/10.1073/pnas.1200844109.

Meyer, Katja M., and Lee R. Kump. 2008. Oceanic Euxinia in Earth History: Causes and Consequences. *Annual Review of Earth and Planetary Sciences* 36 (1): 251–88.

National Aeronautics and Space Administration (NASA). 2020. *Vital Signs of the Planet*. Washington, DC: NASA.

National Snow and Ice Data Center (NSIDC). 2014. *Arctic Sea Ice Continues Low; Antarctic Ice Hits a New High*. Boulder, CO: NSIDC. Accessed from https://nsidc.org/news/newsroom/arctic-sea-ice-continues-low-while-antarctic-reaches-new-record-high.

National Snow and Ice Data Center (NSIDC). 2020. *Sea Ice Index: Arctic-and Antarctic-wide Changes in Sea Ice*. Boulder, CO: NSIDC, University of Colorado.

Oppenheimer, Michael. 1998. Global Warming and the Stability of the West Antarctic Ice Sheet. *Nature* 393: 325–32.

Seo, S. Niggol. 2006. Modeling Farmer Responses to Climate Change: Climate Change Impacts and Adaptations in Livestock Management in Africa. PhD dissertation, Yale University, New Haven, CT.

Seo, S. Niggol. 2016. *Microbehavioral Econometric Methods: Theories, Models, and Applications for the Study of Environmental and Natural Resources*. Amsterdam, CH: Academic Press.

Simpson, R.David, Roger A. Sedjo, and John W. Reid. 1996. Valuing Biodiversity for Use in Pharmaceutical Research. *Journal of Political Economy* 104 (1): 163–85.

TIME Magazine. 2020. *Greta Thunberg: TIME's Person of the Year 2019*. New York, NY: TIME Magazine.

United Nations Framework Convention on Climate Change (UNFCCC). 2015. *Paris Agreement*. New York: UNFCCC.

Weitzman, Martin. 1998. The Noah's Ark Problem. *Econometrica* 66: 1279–98.

Willig, M.R., D.M. Kaufman, and R.D. Stevens. 2003. Latitudinal Gradients of Biodiversity: Pattern, Process, Scale, and Synthesis. *Annual Review of Ecology Evolution and Systematics* 34 (1): 273–309.

CHAPTER 10

A Story of Coral Reefs, Nemo, and Fisheries: On Biodiversity Loss and Mass Extinction

10.1   INTRODUCTION

Last chapter kicked off the discussion on the biodiversity loss and mass extinction that may result from the unfolding of global warming. I led the discussion from the perspective of the overall biological diversity of the Planet and then went on to highlight the plights of the two much-loved animals by the people: polar bears and penguins. I will continue the discussion of the mass species extinction in this chapter with a new focus on marine species and global commercial fisheries.

Of the total 2.2 million marine species, according to one research to be explained shortly, we will have a closer look at the two perhaps most vulnerable marine species widely concerned by people. One is coral reefs and the other is Nemo, a colorful clownfish famed for his film appearance. Both are stunningly beautiful marine animals. Scientists have warned that both may disappear from the Planet owing to changes in the oceans caused by a warming Planet (IPCC 2018).

Another important research area with regard to the ocean and marine species is whether the global fish catch will suffer a great deal because of dying of commercially important fish species caused by changes in the ocean induced by global warming (Cheung et al. 2010). This concerns your daily consumptions of fish including Japanese sushi, Boston clam, and North Sea salmon. More importantly, it concerns the livelihoods of the fishermen across the Planet. Will the fishermen in Thailand, Vietnam,

S. N. Seo, *Climate Change and Economics*,
https://doi.org/10.1007/978-3-030-66680-4_10

171

and West Africa suffer losses of income and nutrition owing to warmer oceans?

The marine species are impacted by climate change because the latter brings about changes in the ocean. The two most noticeable changes are a temperature (heat) increase in seawaters and an ocean acidification from a carbon dioxide uptake from the atmosphere (Johnson et al. 2020; Feely et al. 2020). Scientists have strived to quantify the effects of these changes on marine species via laboratory experiments and enclosed-ocean-area experiments (Kleypas et al. 1999; Gooding et al. 2009). Are marine species particularly vulnerable to the degree that they would die in elevated conditions of the two variables?

Intuitively, we can expect some marine species may be able to swim across the waters to find pleasant places to them, although others such as coral reefs seem to be virtually immobile. This raises the prospect of adaptation of marine species. How different across the species are their adaptive capacity to the aforementioned changes in the oceans? There is an emerging literature on this inquiry, some of which will be discussed in this chapter (Langdon et al. 2018; Salles et al. 2019).

This chapter will also raise an inevitable economic question, which I alluded in the previous chapter: If we cannot preserve all the species of the Planet, which species should be the first ones to be protected (Simpson et al. 1996; Weitzman 1998)? Of course, in the scenario in which there is little change in the total number of species because some species would disappear and other species would newly emerge, we may not need to be too much concerned about this question. Even in this optimistic scenario, however, some citizens may get disturbed if they were to hear some corals or Nemo the clownfish should be left to extinction. The above question remains valid and important even in this optimistic scenario.

In attempting to examine a mass extinction or which species to be conserved, we are led back to the fundamental biological question: After all, how many species of life are there in the land and in the oceans of Earth? An international scientific initiative took up this question recently. Strikingly, only a small fraction of the species on the Planet are named and can be described by scientists up until now, according to the biologists who have pondered over this question for a long time (Mora et al. 2011).

10.2 How Many Marine Species Are on Earth?

When it comes to addressing the question of biodiversity losses and a mass extinction of species, it would be fundamental that we should start with the knowledge of how many species of life are on Earth at the present time. Unfortunately, the human society does not know this number, with only a wide range of estimates offered by the concerned people: 3 million species at a low end to 300 million species at a high end of the range. Even worse, the available estimates are largely from opinions of experts (Mora et al. 2011). As of now, only 1.25 million species are described by biologists and entered into the Linnaean system of classification.

In the last chapter, I introduced to you the biodiversity time-series database, the BioTIME database (Dornelas et al. 2018). This database tracks as many as 50,000 time series of a biodiversity index, some of which goes back to the 1800s. However extensive the database is, it is still a sample of the population of all species, as such, it cannot avoid the problem of selection bias (Mayhew et al. 2012). In other words, it is prone to represent only certain sub-sections of the population of all species.

A correct estimate of the total number of species on the Planet would therefore provide valuable information to the human efforts to preserve biological diversity and avoid a mass extinction. In the chapter on forestry, Chapter 3, I also explained an interesting research effort to quantify the number of trees on the Planet, that is, individual trees not species of trees, which was motivated by the carbon sink potential and the carbon credits to be awarded to the forestry sector (Crowther et al. 2015).

Why is it important for us to know this number at all? To give you a concrete example, according to the International Union of Conservation of Nature (IUCN)' red list of the species threatened with extinction, it has 32,000 such species. The total number of assessed species by the organization is 120,000 (IUCN 2020). The number of assessed species accounts for only 4% of the lower end and 0.04% of the upper end of the aforementioned range of the total number of species, which is too small a fraction of the total population to draw a reasonable conclusion on the top questions of this chapter, that is, a mass extinction. So, depending on which of the two is the correct number of total species, the red list has quite different policy implications.

A recent report by the Census of Marine Life, which was a large world-wide collaborative project, offers us a glimpse into the challenging task of estimating the number of species on the Planet. Relying on the numbers of species in the existing Linnaean system of species classification, the authors estimated that there are 8.74 million eukaryote species on Earth. Of the total, the number of animal species amounts to 7.77 million while the number of plant species to 298,000 (Mora et al. 2011).

In 1735, Carl Linnaeus, a Swedish scientist, created the taxonomy system of species, named *Systema Naturae*, which is still used to name and describe species. It is a pyramid-like hierarchy of all species: from species upwards, genus, family, order, class, phylum, kingdom, and domain on the top (Linnaeus 1758). At present, there are about 1.25 million species named, described, and entered into the catalogue, of which roughly 1 million are land species and 250,000 are marine species.

The Linnaean taxonomy is widely used today and indeed we are quite familiar with it, for example, *Homo Sapiens* that refers to the human being. To give you one of my favorites, the flowering dogwood which is one of the most bright and gracious flowering trees in New England during the springtime is named as follows, with *Cornus* being a genus:

Cornus Florida

From the number of species in the existing catalogue of the Linnaean taxonomy, the authors interpolate to estimate the total number of species on Earth. Of the total 8.74 million species thus estimated, the authors report that 6.5 million species are land species and 2.2 million species are marine species (Mora et al. 2011). Of the total number of marine species estimated to be 2.2 million, which is the primary topic of this chapter, only about 250,000 species (11%) are described by scientists and entered into the catalogue of the Linnaean taxonomy.

The researchers argue that we do not know yet 86% of the land species and 90% of the marine species on the Planet, which have yet to be discovered and described by the humankind, say, *Homo Sapiens*. This is quite striking! Isn't this?

10.3   Can Coral Reefs Survive?

I had the opportunity to visit Manila, the capital city of the Philippines, with an invitation from the Asian Development Bank in around 2007. The next morning of my check in at a hotel, I found the newspaper delivered at the room door with the front-page headline featuring a climate change conference on the Coral Triangle. The Coral Triangle is a roughly triangular area in the tropical seas of the Philippines, Malaysia, Indonesia, Papua New Guinea, and Solomon Islands.

I also had the opportunity to teach in Sydney, Australia for about five years, another hotspot of coral reefs. The Great Barrier Reef is located in the Coral Sea off the coast of the State of Queensland. It is the world's largest coral reef system, boasting about 3,000 individual reefs and over 900 coral islands called cays.

In the winter following my visit to Manila, if my memory is correct, I heard from Stephen Schneider, a pioneering climate scientist from Stanford University, at a Yale University conference on climate change that the coral reef is the evidence of a local species extinction that we can observe currently. He said this in a response to the question, posed by none other than William Nordhaus, whether there is a scientific evidence of global warming-caused mass extinction of species.

Getting back to the present, let's start our analysis of a marine species extinction with coral reefs. In fact, the Intergovernmental Panel on Climate Change (IPCC) issued a dire warning on the threats faced by coral reefs from climate change. It predicts with high confidence that 70–90% of coral reefs will disappear under a 1.5 °C of warming and nearly all of them will disappear with a 2 °C warming (IPCC 2018).

The coral reefs have received the most attention of all marine species because of, among other reasons, their natural beauty under the seas and their tourist attractions. Corals are also peculiarly an animal, a vegetable, and a mineral all at once. The dire warning by scientists that a continued Planet warming will threaten the survival of coral reefs is rooted in the ocean acidification which reduces calcification rates of corals (Kleypas et al. 1999). The ocean acidification is caused by a net carbon uptake of the oceans from the atmosphere which results in an increase of carbon dioxide in the oceans (Lumpkin 2020).

Coral bleaching, on the other hand, is caused by warmer seawater temperatures which result from the ocean heat uptake from the atmosphere (Johnson et al. 2020). A warmer seawater makes corals to expel the algae living in them, as a result of which the once colorful coral reefs will turn completely white. A big coral bleaching event around the globe occurs in a strong El Nino year, *e.g.*, the 1997–1998 event and the 2015–2016 event, when the parts of the Pacific Ocean heat up.

Even if bleached, however, the corals are not dead. They can recover quickly when the waters around them become cooler again. A sustained cooling period after the El Nino years, for example, would lead to a recovery of the corals' colors and beauty. If ocean temperatures were to be increased "permanently" by global warming, on the other hand, the bleached corals could not recover unless they could find ways to adapt to higher temperatures.

A dying of corals caused by the ocean acidification is perhaps a greater threat to the coral reef community. A recent experiment at the southern Great Barrier Reef tried to restore the seawater condition to the pre-industrial times by reducing acidity of the water, that is, by adding alkalinity. The authors reported a 7% increase in coral reef calcifications by restoring the seawater condition (Albright et al. 2016).

You might conclude from this experiment that humanity can stop the dying of coral reefs from the acidifying oceans. This is a hasty conclusion, although not groundless. The experiment was successful in a particular location of the ocean with three lagoons, a favorable site for the experiment of controlling the acidity in closed waters. Increasing alkalinity in the oceans at a global scale or in an open ocean would be by far more difficult and would possibly cause significant side effects.

Can coral reefs adapt to warmer seawaters as well as ocean acidification? The literature on this question seems to be emerging rapidly. A recent experiment compares two threatened coral species found in Florida and the Caribbean Sea to identify their differences in adaptability to high temperatures and ocean acidification. One of them, the mountainous star coral, survived in an elevated temperature condition from 26 °C to 32 °C and also in an elevated acidity environment to a certain level. The other one, the staghorn coral, survived in an elevated acidity environment from 380 ppm to 800 ppm of carbon dioxide, but not in an elevated temperature (Langdon et al. 2018).

The study indicates that a variety of coral species have varied adaptive capacities to additional stresses and more resilient corals to them will expand while less resilient corals will shrink. The authors argue that the mountainous star corals, which have special algae living in them, will be able to survive under the two additional climate change-induced stresses by the end of this century, that is, higher temperatures and elevated acidity.

10.4   Fish Catch and Finding Nemo in Colder Seas?

The coral reefs are only one of the marine species, 2.2 million of them in total according to the above-referenced modeling. Of all the marine species, some species are well known to humans while many others are not even identified even by the experts. Of the former, some species are commonly known to humanity for their importance as food sources while others gained popularity through a popular culture, to give you just one example, Nemo in the blockbuster movie "Finding Nemo."

How many fish does the human society catch and consume each year? The total production of fish in 2018 was 180 million tons globally. This can be decomposed into 84 million tons from marine fisheries, 12 million from inland capture fisheries, 31 million from marine aquaculture, and 51 million from freshwater aquaculture. Noticeably, the fish catch from aquaculture, which is fish farming, is as large as the catch from the oceans and seas. Each person on the Planet consumed 20.5 kg of fish in that year on average (FAO 2020).

Fish is one of the primary sources of food and nutrition to humanity. It provides 17% of the animal protein consumed by humankind and as much as 50% of it in many least-developed countries (UN 2017). As such, an important food security question is whether the fish production of the Planet will be harmed, severely or modestly, by an unfolding of global warming.

Figure 10.1 gives us a clue on the answer, which shows the changes in the world's annual total fish catch of the four types of fish since the early 1960s to the early 2010s: demersal (living near the bottom of seas), pelagic (living neither close to the bottom nor near the shore), freshwater (living in lakes and rivers), and marine-other. The total fish catch has increased over the five-decade period across all four types of fish, especially markedly for freshwater fish and pelagic fish (FAO 2020). Do not miss that the increase in fish catch occurred despite the global warming realization over this period.

To get a closer look, I draw in Fig. 10.2 the changes in the annual total fish catch for each of the world major fishing regions. The total catch is defined as the sum of the catches of the above-mentioned four types of fish. The increase in fish catch in Southeast Asia is remarkable, nearly a tenfold increase over the five decades. The other regions where the fish catch is shown to have increased notably are Africa and South America.

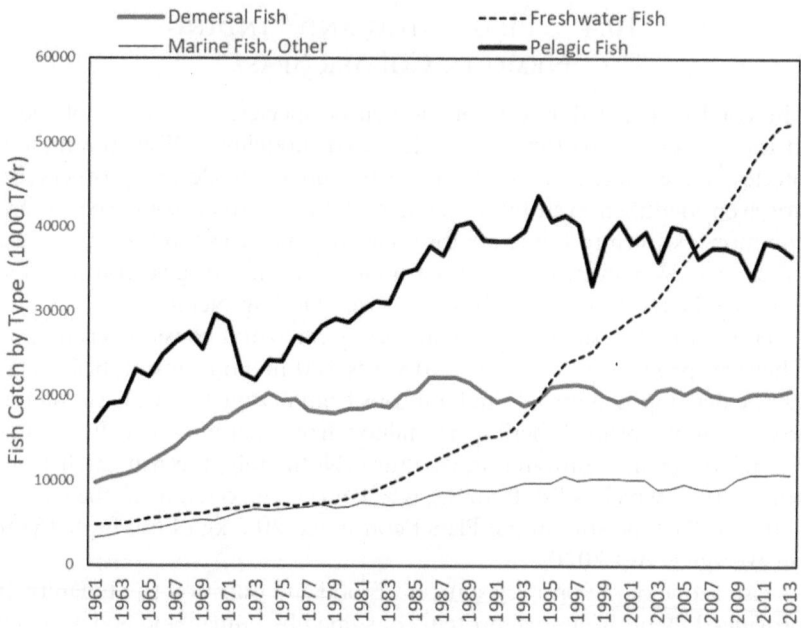

Fig. 10.1   The total fish catch by fish type (1000 tons/year)

On the other side, the decrease in Japanese fish catch is striking since the late 1980s, which is likely attributable to international regulations (FAO 2020).

If you are a gourmet of Japanese sushi, Boston clam, or North Sea salmon, it will not be hard for you to realize that many fish species may prefer cold waters to warm waters. If they do better in colder waters, would a warmer ocean decrease the number of fish in the oceans? On second thought, however, you may quickly get that Nemo, a vibrantly colorful clownfish, will not hesitate to swim all the way down from the Great Barrier Reef to the city of Sydney if he would find the warm waters unpleasant. The point is that some fish, if not most, can swim a long way across the seawaters, unlike a water buffalo in India or a corn plant on the field of Des Moines in Iowa.

Considering their mobility, it is not contradictory to our intuition that a scientific research predicts little change in the global fishery productivity owing to global warming but significant changes in the distributions of

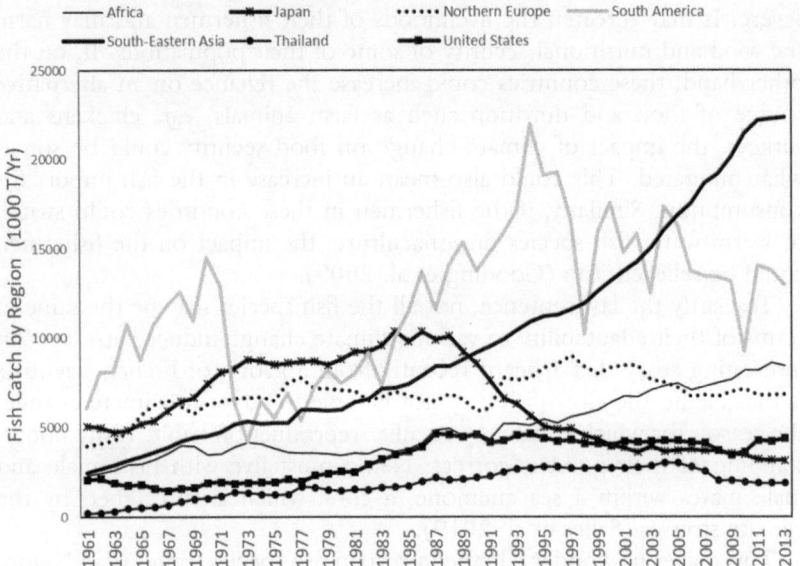

Fig. 10.2   The total fish catch by fishing region (1000 tons/year)

fish species across the global oceans (Porter et al. 2014). To be more specific, the research predicts a 30–70% increase in fish catch potential in high latitude regions and a decrease of up to 40% in the tropical regions. The authors relied on the maximum catch potential model and examined 1,066 commercially-exploited fish species, covering about 70% of all such species (Cheung et al. 2010).

Interestingly, the life of the fish-loving gourmet may not be affected much. According to the above model, the maximum fish catch is predicted to increase in Japanese fishing areas, New England coasts in the North Atlantic, and the North Sea. By contrast, the maximum fish catch potential is predicted by the modelers to decline in warmer waters of Thailand, other South Asian countries, and West Africa, which would be concerning to these countries and fishermen thereof (Cheung et al. 2010).

Considering the importance of the fishery in the latter countries as a food supplier and an employer of their populations (FAO 2020), the impacts of climate change on the fisheries and fishermen there may be severe. It may threaten the livelihoods of their fishermen and may harm the food and nutritional security of some of their populations. If, on the other hand, these countries could increase the reliance on an alternative source of food and nutrition such as farm animals, *e.g.*, chickens and turkeys, the impact of climate change on food security could be somewhat mitigated. This could also mean an increase in the fish import for consumption. Similarly, if the fishermen in these countries could switch to warm-water fish species or aquaculture, the impact on the fishermen could be relieved, too (Gooding et al. 2009).

To clarify the last sentence, not all the fish species may be the same in terms of their adaptability to various climate change-induced stresses. An interesting story that I heard recently from a group of French scientists is that Nemo the clownfish may not be able to adapt to climate change, the reason for which is his very peculiar reproduction habit. If my understanding of their point is correct, Nemo must live with his female and male mates within a sea anemone, a coral which is threatened by the climate stresses (Salles et al. 2019).

This makes us wonder whether a certain fish species, other than Nemo, can adapt to the new ocean environments to be brought about by a climatic shift. The evidence in support of this seems to be emerging. A research reports that the winter skate fish off Canadian coasts simply switches on certain genes in them to survive in warmer waters (Lighten et al. 2016). For another, an experiment on starfish shows that the growth of the fish is enhanced under a hotter water temperature as well as under increased CO_2 conditions in the water (Gooding et al. 2009). Considering that these experiments are set up for the responses of the concerned fish species within several weeks, we can expect that many fish species may be able to adapt to the oceans that are getting warmer in an incremental slow-paced manner.

10.5   Which Species to Be Preserved or Let Go to Extinction?

From this point on, let's shift our focus to the policy question of conservation of species, which was first introduced in the last chapter (CBD 1992, UNFCCC 2015). Of the whole array of endangered and vulnerable marine and land species under a warmer Planet that are under consideration for preservation by policy-makers and biologists, which ones should be the first to be preserved or given priority in the preservation efforts

(Weitzman 1998)? The array of threatened species is expansive, possibly containing nearly a million species and including such people's favorites as a monarch butterfly, Nemo the clownfish, coral reefs, polar bears, and penguins (IUCN 2020). As such, a systematic framework for the selection decision and the funding decisions is called for, rather than a case-by-case haphazard approach.

In the sound framework, a number of variables should be carefully weighed against each other, including the following ones. First, how strongly people want to preserve the species under review? Second, how much could the conservation program under consideration increase the survival probability of the endangered species? Third, what is the expected cost of the conservation program of the species? Fourth, how unique or distinctive the endangered species is in the ecosystem? Fifth, how likely is the endangered species to adapt and evolve under a warmer climate?

A systematic framework proposed by Martin Weitzman attempts to achieve this by balancing these considerations for the species conservation efforts from the perspective of the society's welfare optimization. According to his approach, the ranking of an individual species conservation project (i) from the whole list of projects (I) should be determined by the following index (Weitzman 1998):

$$R_i = (U_i + D_i) * \left(\frac{\Delta P_i}{C_i} \right), i = 1, 2, \ldots, I. \qquad (10.1)$$

The Noah's Ark problem, as Eq. 10.1 was referred to as by Weitzman himself, provides a ranking (R_i) of an individual species preservation program of all programs under consideration. This is similar to the quandary of Noah, as such, an apt metaphor for the conservation decision-making.

To clarify Eq. 10.1, the ranking is given on the basis of the "expected marginal distinctiveness plus utility per dollar" spent on the program. The utility is U_i. The uniqueness or distinctiveness of the species in the Planet's ecosystem is measured by the distance from other species, D_i, to be clarified presently. The increase in the probability of survival of the species owing to the conservation effort is ΔP_i. The cost of the program is denoted by C_i.

As succinct as the equation is, the quantification of R_i is not easy at all. It involves advanced science and economics. For a starter, the utility in Eq. 10.1 is the sum of the willingness to pay of an individual for the conservation of the species across all individuals. How such a value can be drawn from an individual's actions or statements is arguably one of the most challenging tasks in economics (Simpson et al. 1996; Mendelsohn and Olmstead 2009). The former is referred to as a revealed preference method of valuation. Most commonly, researchers ask each surveyed individual to state directly the monetary value s/he is willing to pay (sacrifice) for the conservation of the species, which is referred to as a stated preference method of valuation (Hanemann 1994). The stated preference method could, however, result in the biases that are well established in the valuation literature, inter alia, giving strategic answers.

The distinctiveness (D_i) of species i from the "library" of biodiversity should be determined only from a vast scientific endeavor. It could very well involve comparisons of genetic information of all species in the basket of species under consideration for conservation. The less overlap there is in the genetic information of the species with those of the other species, the higher the distinctiveness of the species is. This task is far amplified owing to the absence of the complete library of species, as discussed in Sect. 10.2 of this chapter (Mora et al. 2011).

For the ranking, an increase in the survival probability of each species through the conservation dollars spent should also be estimated. The survival probability would get improved more per each dollar spent if the species would have a higher adaptation capability. Further, other than the survival of a species itself, people would be similarly concerned about a decrease in the population of a (favorite) species or an increase in the population of an (invasive) species.

10.6   Final Words: The Resilient Oceans

This chapter is a sequel of the preceding chapter on a mass extinction of species and biodiversity losses, with a fresh focus on marine species. Of the marine species, people continue to have a grave concern on the health of coral reefs. To some degree, the review of literature in this chapter offers a relief: The corals may be able to bounce back from bleaching events and some corals seem to be able to adapt to higher seawater temperatures better.

As far as the commercial fisheries are concerned, the total fish catch at the global level may not suffer a great loss because of a warming Planet. However, some regions and the fishermen thereof may suffer significantly, for example, Southeast Asian fisheries. The ultimate impacts, however, will depend on how fishermen can adapt their practices as well as whether fish species can develop adaptive capacities to warmer oceans.

On the balance of the scientific evidence presented up until now, it seems premature to be worried too much about the mass extinction of species on the Planet. Rather, it points us to the necessity of a rational conservation decision-making: Which species should be protected first of all the vulnerable species and how? This is a fascinating inquiry to pursue which entails for a satisfactory decision framework a multi-disciplinary endeavor involving science, economics, ecology, ethics, and value.

Before leaving this chapter behind which has concentrated on the impacts of global warming on the marine species in the oceans, I would like to remind the readers of the high capacity of the world oceans. The oceans are not fragile, which I explained in Chapter 5 via the atmosphere ocean carbon exchange as well as the atmosphere ocean heat exchange. The oceans absorb carbon dioxide from the atmosphere: The ocean carbon uptake has been increasing to roughly 2.5 GtC or PgC per year, which is roughly a quarter of the annual anthropogenic emissions of carbon from fossil fuels (Feely et al. 2020).

As per the ocean heat uptake, the oceans absorb the heat in the atmosphere. Without the oceans, the Planet would be a lot hotter today. Of the total amount of heat energy added to the Planet from the global warming during 1971–2010, over 90% of the heat is absorbed by the world oceans (Johnson et al. 2020).

10.7   Chapter Highlights

- This chapter is a sequel of the preceding chapter on a mass extinction of species with a fresh focus on marine species and the oceans.
- There is a wide range of estimates on the total number of species on Earth. A new scientific research based on the Linnaean taxonomy puts it at about 8.8 million species of which 2.2 million are marine species.
- A global total fish catch may change little owing to a warming ocean, but some regions may suffer a great loss in their fisheries. A study estimates that the fish catch will increase by as much as 70% in the high latitude cold oceans while it will decrease by 40% in the tropical warm oceans.
- Coral reefs and some marine species such as Nemo the clownfish are predicted to be more vulnerable because of their immobility or a

special reproductive habit. Some fish species are, however, reported to have higher adaptive capabilities to both high water temperatures and increased CO_2 levels.

- Corals are, however, observed to recover quickly from bleaching events. Recent studies also indicate that there are some types of coral species that are resilient to both high water temperatures and ocean acidification.

- Which species should be preserved while others should not if we cannot protect all vulnerable species? The Noah's ark framework, based on an apt metaphor, offers a rational conservation decision framework from the viewpoint of an economic optimization.

REFERENCES

Albright, Rebecca, Lilian Caldeira, Jessica Hosfelt, Lester Kwiatkowski, Jana K. Maclaren, Benjamin M. Mason, Yana Nebuchina, et al. 2016. Reversal of Ocean Acidification Enhances Net Coral Reef Calcification. *Nature* 531 (7594): 362–65.

Cheung, William W.L., Vicky W.Y. Lam, Jorge L. Sarmiento, Kelly Kearney, Reg Watson, Dirk Zeller, and Daniel Pauly. 2010. Large-Scale Redistribution of Maximum Fisheries Catch Potential in the Global Ocean under Climate Change. *Global Change Biology* 16 (1): 24–35.

Convention on Biological Diversity (CBD). 1992. *Convention on Biological Diversity*. New York, NY: United Nations.

Crowther, Thomas W., Henry B. Glick, Kristofer R. Covey, C. Bettigole, D.S. Maynard, S.M. Thomas, J.R. Smith, et al. 2015. Mapping Tree Density at a Global Scale. *Nature* 525 (7568): 201–5.

Dornelas, Maria, Laura H. Antão, Faye Moyes, Amanda E. Bates, Anne E. Magurran, et al. 2018. BioTIME: A Database of Biodiversity Time Series for the Anthropocene. *Global Ecology and Biogeography* 27: 760–86. https://doi.org/10.1111/geb.12729.

Feely, R.A., R. Wanninkhof, P. Landschützer, B.R. Carter, and J.A. Triñanes. 2020. Global Ocean Carbon Cycle [in State of the Climate in 2019]. *Bulletin of the American Meteorological Society* 101 (8): S170–75.

Food and Agriculture Organization (FAO). 2020. *The State of World Fisheries and Aquaculture 2020*. Rome, IT: FAO.

Gooding, Rebecca A., Christopher D.G. Harley, and Emily Tang. 2009. Elevated Water Temperature and Carbon Dioxide Concentration Increase the Growth of a Keystone Echinoderm. *Proceedings of the National Academy of the Sciences* 106 (23): 9316–21. https://doi.org/10.1073/pnas.0811143106.

Hanemann, W.Michael. 1994. Valuing the Environment Through Contingent Valuation. *Journal of Economic Perspectives* 8 (4): 19–43.

Intergovernmental Panel on Climate Change (IPCC). 2018. *Special Report on Global Warming of 1.5 °C*. Cambridge: Cambridge University Press.

International Union for the Conservation of Nature (IUCN). 2020. *The IUCN Red List of Threatened Species*. Version 2020-2. Gland, CH: IUCN, 2020. Accessed from https://www.iucnredlist.org.

Johnson, G.C., J.M. Lyman, T. Boyer, L. Cheng, C.M. Domingues, J. Gilson, M. Ishii, R.E. Killick, D. Monselesan, S.G. Purkey, and S.E. Wijffels. 2020. Ocean Heat Content [in State of the Climate in 2019]. *Bulletin of the American Meteorological Society* 101 (8): S140–44. https://doi.org/10.1175/BAMS-D-20-0105.1.

Kleypas, Joan A., Robert W. Buddemeier, David Archer, Jean-Pierre Gattuso, Chris Langdon, and Bradley N. Opdyke. 1999. Geochemical Consequences of Increased Atmospheric Carbon Dioxide on Coral Reefs. *Science* 284: 118–20.

Langdon, Chris, Rebecca Albright, Andrew C. Baker, Paul Jones. 2018. Two Threatened Caribbean Coral Species Have Contrasting Responses to Combined Temperature and Acidification Stress. *Limnology and Oceanography*, July 31. http://doi.org/10.1002/Ino.10952.

Lighten, Jackie, Danny Incarnato, Ben J. Ward, Cock van Oosterhout, Ian Bradbury, Mark Hanson, and Paul Bentzen. 2016. Adaptive Phenotypic Response to Climate Enabled by Epigenetics in a K-Strategy Species, the Fish Leucoraja Ocellata (Rajidae). *Royal Society Open Science* 3 (10): 160299. https://doi.org/10.1098/rsos.160299.

Linnaeus, Carolus. 1758. *Systema naturæ per regna tria naturæ, secundum classes, ordines, genera, species, cum characteribus, differentiis, synonymis, locis*. Stockholm, SE: Laurentius Salvius.

Lumpkin, R.L. (ed.). 2020. Global Oceans [in "State of the Climate in 2019"]. *Bulletin of the American Meteorological Society* 101 (8): S129–83. Accessed from https://doi.org/10.1175/BAMS-D-20-0105.1.

Mayhew, Peter J., Mark A. Bell, Timothy G. Benton, and Alistair J. McGowan. 2012. Biodiversity Tracks Temperature over Time. *Proceedings of the National Academy of Sciences* 109 (38): 15141–45.

Mendelsohn, Robert, and Sheila Olmstead. 2009. The Economic Valuation of Environmental Amenities and Disamenities: Methods and Applications. *Annual Review of Environment and Resources* 34 (1): 325–47.

Mora, Camilo, Derek P. Tittensor, Sina Adl, Alastair G.B. Simpson, and Boris Worm. 2011. How Many Species Are There on Earth and in the Ocean? *PLoS Biology* 9 (8): e1001127. https://doi.org/10.1371/journal.pbio.1001127.

Porter, John R., L. Xie, A.J. Challinor, K. Cochrane, S.M. Howden, M.M. Iqbal, D.B. Lobell, and M.I. Travasso. 2014. Food Security and Food Production

Systems. In *Climate Change 2014: Impacts, Adaptation, and Vulnerability*, ed. C.B. Field et al. Cambridge: Cambridge University Press.

Salles, Océane C., Glenn R. Almany, Michael L. Berumen, Geoffrey P. Jones, Pablo Saenz-Agudelo, Maya Srinivasan, Simon R. Thorrold, Benoit Pujol, and Serge Planes. 2019. Strong Habitat and Weak Genetic Effects Shape the Lifetime Reproductive Success in a Wild Clownfish Population. *Ecology Letters*. https://doi.org/10.1111/ele.13428.

Simpson, R.David, Roger A. Sedjo, and John W. Reid. 1996. Valuing Biodiversity for Use in Pharmaceutical Research. *Journal of Political Economy* 104 (1): 163–85.

United Nations (UN). 2017. Our Ocean, Our Future: Call for Action. Resolution Adopted by the General Assembly on 6 July 2017. The Ocean Conference. New York, NY: UN.

United Nations Framework Convention on Climate Change (UNFCCC). 2015. *Paris Agreement*. New York: UNFCCC.

Weitzman, Martin L. 1998. The Noah's Ark Problem. *Econometrica* 66 (6): 1279–98.

CHAPTER 11

A Story of Infectious Diseases and Pandemics: Will Climate Change Increase Deadly Viruses?

11.1 THE NOVEL CORONAVIRUS AND PANDEMICS

At the time of this writing in April 2020, the world is in the deep of the worldwide pandemic, the novel coronavirus. Unleased from Wuhan, China at the end of 2019, the virus spread to Asian countries, Europe, and the US. As of April 2020, over 2.5 million people worldwide were tested positive and about 180,000 people lost lives. Many States in the US were placed into lockdowns since the middle of March and nearly all businesses were forced to be closed. By the middle of August 2020, the number of people infected passed 22.3 million and the total fatality is near 800,000 worldwide (JHU 2020).

The contagiousness of the novel coronavirus is reported to be unprecedented, even compared with the recent outbreaks of the infectious diseases such as H1N1 influenza, MERS, and SARS virus. Besides these recent virus outbreaks, the human society has long fought deadly viruses to survive, to name the deadliest ones during the past century, Spanish flu in 1918, Asian flu in 1957, small pox, and measles (Patz et al. 2003; Henderson et al. 2009). These diseases are an infectious disease transmissible from human to human, in scientific terms, *anthroponoses*. Many scientists argue that the deadly viruses are the greatest challenge faced by humankind, that is, not climate change nor a nuclear war.

On the other hand, there are some who argue that climate change and fatal viruses are not mutually exclusive. Specifically, they argue that

S. N. Seo, *Climate Change and Economics*, https://doi.org/10.1007/978-3-030-66680-4_11

climate change may increase the occurrences of fatal viruses and further make them more contagious and fatal. An often-cited example is malaria, still existent in Africa and India, the two hottest continents on Earth. They argue that malaria may become even more contagious, spreading to the other continents, and more fatal owing to a Planetary warming.

Of the infectious diseases that affect humanity seriously, some are carried from one animal to another, *zoonoses* in scientific terms, which could be a valuable asset to many people. It is often carried by a vector, say, mosquitoes and flies. Of the most well-studied livestock diseases are sleeping sickness in African cattle, *trypanosomiasis* in scientific names, and blue tongue in sheep (Ford and Katondo 1977). In a severe outbreak year, the *trypanosomiasis* may kill millions of cattle in Sub-Saharan countries as well as thousands of African people.

The sleeping sickness, also commonly called Nagana, is caused by infection with protozoan parasites. A tsetse fly carries the infectious agent, also called pathogen, from cattle to cattle, so it is called a carrier or a vector. For the vector-borne diseases such as Nagana, we should be interested in understanding whether the disease-carrying vectors and the parasites are affected by changes in the climate system (Aksoy et al. 2014). Will the tsetse flies become more prevalent or the parasites become more powerful owing to a warming Planet?

This chapter will review the range of topics in the literature of infectious diseases and climate change which have received much attention by the general public. It will introduce the major infectious diseases, mostly the major pandemics, since the dawn of the twentieth century, although most of them may not have been climate change-caused. What the present author wants to accomplish in this chapter is, among other things, to separate scientific facts from the fictional stories which are intended solely to bring fear to the public.

It is notable that fear is recognized as an important policy option in many countries' efforts to control the spread of an infectious disease among their populations, whose manifestation can be seen in a nationwide lockdown, a crackdown on a certain group of people, a universal mask mandate, and a GPS monitoring of every citizen. A fear policy could have an undesirable outcome of ending up to be clouding the scientific data which is the foremost weapon for the fight against the pandemics.

I hope this chapter will help you to assess on your own whether the world needs to be terrified of the future viruses and infectious diseases that could be caused by a future climatic system.

11.2 Climate Change and Infectious Diseases

Infectious agents are microbes that cause infectious diseases, which include viruses, bacteria, protozoa, and multicellular parasites. The infectious agents, scientifically called the pathogens, can be categorized into the agents that cause either *anthroponoses* (human-to-human diseases) or *zoonoses* (animal-to-animal diseases). The microbes that cause *anthroponoses* have adapted to the human species as their exclusive host.

The list of infectious diseases can be classified as either directly-transmitted diseases or indirectly-transmitted diseases. The former includes TB (Tuberculosis), HIV/AIDS, measles for the human-to-human *anthroponoses*; and rabies for animal-to-animal *zoonoses*. The latter includes malaria, dengue fever, yellow fever for the human-to-human *anthroponoses*; Bubonic plague, Lyme disease for the animal-to-animal *zoonoses*.

The indirectly-transmitted diseases are carried by a disease vector, more commonly a disease carrier, which includes mosquitoes, flies, lice, ticks, and even dust particles. The infectious diseases transmitted by vectors are called the vector-borne diseases (Patz et al. 2003).

An analysis of the impacts of climate change on infectious diseases can be decomposed into three components: vectors, pathogens, and hosts. The vectors, pathogens, and hosts are dependent and survive in the range of climatic conditions. Temperature and precipitation are the most critical climate variables but other climate variables such as daylight duration and wind are also important.

The most common infectious disease that attacks humans is influenza, commonly known as the flu. Of the influenza viruses, influenza A virus subtype H1N1 is the deadliest. It was the most common cause of the two pandemics in human history: the 2009 H1N1 swine flu pandemic killing 284,000 people and the Spanish flu pandemic in 1918 killing up to 50 million people.

At the time of this writing, the novel coronavirus, called COVID-19, is afflicting the human society immensely, with nearly 800,000 human deaths. The novel coronavirus in 2020 is a new strain of the SARS (Severe Acute Respiratory Syndrome) coronavirus in 2002–2004. The MERS (Middle East Respiratory Syndrome) coronavirus spread in South Korea in 2015 and in Saudi Arabia in 2018 (JHU 2020).

How can scientists predict an outbreak, contagiousness, and fatalities from a future infectious disease? It is a monumental task because the viruses and other pathogens are evolving and adapting to find a weak spot in the human and animal bodily functions. The SARS coronavirus evolved to the MERS coronavirus, and then to the novel coronavirus, COVID-19. How it will evolve is unpredictable at the current status of virology and medical science.

What about the already existing viruses and diseases? The scientific task may become a little more manageable. Scientists employ three scientific methods to predict changes in the infectious diseases: a statistical model, a process-based model, and a landscape-based model (Patz et al. 2003).

To our dismay, the three approaches yield contrary predictions. Statistical models indicate little changes in climate change-induced infectious disease outbreaks (Rogers and Randolph 2000). Process-based models indicate otherwise. Specifically, they predict that, a temperature increase of 2–3 °C would increase the number of people who are at risk of malaria by around 3–5% globally, that is, by several hundred million people (Martens et al. 1999). Although we cannot draw a definitive conclusion from the three approaches, we still have another option to proceed to tackle the question under consideration. We can examine the individual infectious diseases one by one which turned out to be most fatal in order to assess whether there is any evidence that one of these diseases will be even more fatal under a warmer Planet.

11.3   SMALLPOX: A SUCCESS STORY
OF ERADICATION OF A DEADLY VIRUS

The smallpox was one of the most feared and deadly infectious diseases in human history until 1980 at which time the virus was declared to be eradicated from the Planet. The present author also grew up with a fear of the virus among the family members during the 1970s, which unfailingly spread from one person to another, but there were vaccines widely available at the time. The smallpox virus was highly contagious and at the same time deadly, killing 3 out of 10 persons infected.

The first case or description of the smallpox disease may have occurred over two thousand years ago in history: third century BCE in the mummies in Egypt or fourth century CE in China. The smallpox virus spread across the Planet through trades and wars. The last case of the smallpox in humanity occurred in 1978 and the World Health Organization (WHO) declared a complete eradication of the smallpox in 1980 after global eradication campaigns (CDC 2020a).

The eradication of the smallpox virus is regarded as the greatest success story in the humanity's fight against deadly viruses. How was it accomplished? What does it mean for global warming researchers? The achievement was made possible in part through the worldwide mitigation programs but largely owed to the advances of the vaccination. The vaccination is likely the key to the future fights against pandemics, so needs clarifications.

The smallpox is caused by a variola virus, of which there are variola major and variola minor. The control efforts of smallpox began from variolation and were completed with vaccination. The variolation is the process by which a material from smallpox sores was given to people who had never had a smallpox, hoping for the recipient to develop the immunity against a naturally occurring smallpox. It is a primitive form of vaccination, which is still important in the absence of a vaccine, e.g., the use of blood plasma from the COVID-19 recovered patients.

The possibility of vaccination surfaced in 1796 when an English Doctor Edward Jenner discovered that the milkmaids who had gotten the cowpox virus did not develop any symptoms of smallpox after variolation, which was published in 1801 on the title "On the Origin of Vaccine Inoculation (Jenner 1801)."

The vaccination, or interchangeably the immunization, refers to the administration of a vaccine to help the immune system develop protection from the virus, whose principle was established by Louis Pasteur through laboratory-based microbiology works (Schwartz 2001). The smallpox vaccine, named the Dryvax, played a critical, if not dominant, role in the global eradication of the smallpox (AP 2008). Today, variola viruses all disappeared while remnant stocks are stored in two World Health Organization (WHO) collaborating centers for research purposes: one at the Center for Disease Control (CDC) in Atlanta in the US and another at the State Research Center of Virology and Biotechnology in Koltsovo, Russia.

The vaccination, however, has a key shortcoming in the fight against the deadly viruses, which is heartfelt by the world citizens as of today under the global spread of COVID-19. It takes many years to develop a vaccine. Once the deadly virus such as the novel coronavirus (COVID-19) in 2020 is created, it becomes highly contagious. In a span of a week, it can spread to millions of people, many of whom will eventually die. The virus is a new microbe "unseen" before by the humanity and scientists. As such, it takes long time to develop a vaccine, four years on average, and manufacture it on a massive scale.

To give you an idea, in the three-month period from the declaration of the COVID-19 to be transmissible from a human being to another human being, the virus had already killed nearly 200,000 people worldwide (JHU 2020). In six months after the declaration, the human fatality is near 800,000. Before a vaccine to become available to the public, humanity needs to rely on prevention and treatments.

11.4   ANTHROPONOSES FOR HUMANS: FLU AND MALARIA

As far as the impacts of climate change on infectious diseases that are transmitted from human to human are concerned, two diseases have received the most attention by climate researchers: influenza (flu) and malaria. The two are saliently associated with climatic conditions. The former occurs in cold seasons in the Northern Hemisphere while the latter is prevalent in tropical climate zones. As such, a shift in the climate system is predicted to lead to changes in occurrences, contagiousness, and fatality of these diseases.

Malaria is an infectious disease caused by a parasite which infects a certain type of mosquito which feeds on humans. Today, it is a public health problem in Sub-Saharan Africa and India. As Table 11.1 shows, the number of deaths owing to malaria was 1.208 million people worldwide during the 1990–2001 time period, which accounted for 2.1% of all disease deaths (Refer to WHO 2019 for the recent estimates). The burden of malaria is not even at all across the world, however. Nearly

Table 11.1 Mortality and the burden of Malaria by World Region from 1990 to 2001

	Low- and middle-income countries		High-income countries		World	
	Deaths	DALYs	Deaths	DALYs	Deaths	DALYs
All causes (in 1000s)	48,351	48,351	7,891	149,161	56,242	1,535,871
Malaria (in 1000s)	1,207	39.961	0	9	1,208	39,970
Malaria (in % of the total causes)	2.5%	2.9%	0	<1%	2.1%	2.6%

Note (1) DALY (Disability-Adjusted Life Years); (2) Modified from Lopez et al. (2006)

all malaria-caused deaths occurred in low- and middle-income countries. There was no malaria-caused death in high-income countries. A similar conclusion can be drawn from the Disability-Adjusted Life Years (DALY) statistic which is a measure of disease burden calculated as the number of life years lost due to early death, ill health, or disability (Lopez et al. 2006).

If climate change were to turn out to make Earth hotter, the number of deaths as well as the number of people at risk of malaria would increase worldwide if we believed in the aforementioned process-based models. On the other hand, Table 11.1 indicates that the number of malaria-caused deaths will decline if low- and middle-income countries should develop economically in the twenty-first century. From this point of view, notice a particularly fast rate of economic growth achieved in India in the recent decades, a country where malaria is still extant.

In predicting the number of future malaria-caused deaths, we need to take into account the pivotal parameter: availability of a vaccine. As long as there is a vaccine for the infectious disease, an outbreak as well as a national-scale spread can be contained or prevented with a thoughtfully-designed disease control strategy.

According to the World Health Organization (WHO), RTS, S/AS01 is the vaccine candidate, developed with the financial support from the Bill and Melinda Gates Foundation, which recently completed its clinical trial of 15,460 children in seven Sub-Saharan countries beginning from 2009. The results received a positive scientific opinion from the European Medicines Agency (WHO 2020).

Another human-to-human transmitted infectious disease that is prominently linked to the climate system is influenza. In particular, the flu virus spreads in a cold climate condition. The number of flu deaths during the past decade in the US is summarized in Table 11.2. A single flu season results in the deaths of 12,000 to 61,000 people in the US, the hospitalizations of 140,000 to 810,000 people in the US, and the deaths of 291,000–645,000 people globally (CDC 2020b; Iuliano et al. 2018).

Since the flu virus survives in the cold weather, that is, the flu season, an increase in the flu season temperature caused by a shift in the climate system is expected to decrease the number of deaths resulting from the flu each year. In this regard, Fig. 11.1 is pertinent, which shows the relationship between mortality rate (per 100,000 people) and the average daily temperature using the US data. The figure shows a U-shape response, meaning that the number of deaths declines as the daily average temperature becomes higher. The figure shows the trough at 60–69 °F which is

Table 11.2 Estimated Influenza (Flu) disease burden in the US

Season	Hospitalizations (in 1000s)	Deaths (in 1000s)
2010–2011	290	37
2011–2012	140	12
2012–2013	570	43
2013–2014	350	38
2014–2015	590	51
2015–2016	280	23
2016–2017	500	38
2017–2018	810	61
2018–2019	490	34
2019–2020, preliminary	[410–740]	[24–62]

Modified from CDC (2020b)

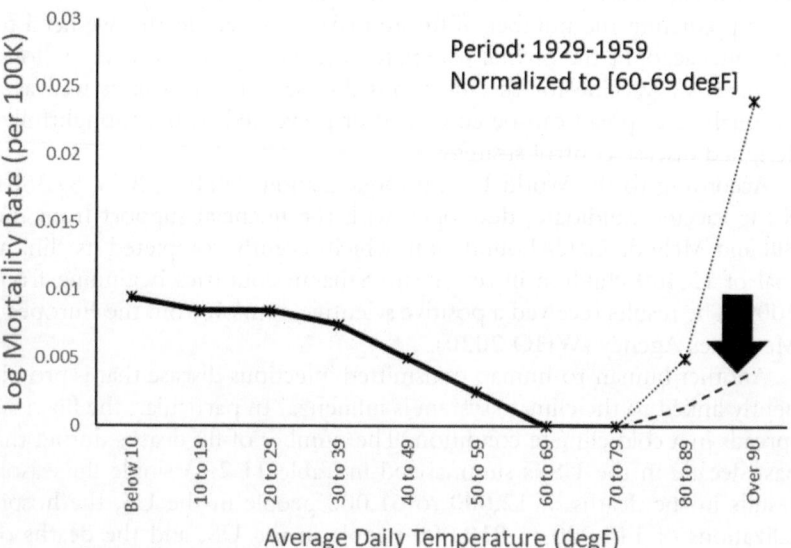

Fig. 11.1 Relationship between mortality and cold temperature in the US (*Note* Drawn from the data at Barreca et al. [2016])

about 15.5–20.5 degree in Celsius (Barreca et al. 2016). Note that the figure is drawn from the data for the 1929–1959 time period.

The dotted line in the figure is in the temperature range that exceeds 70 °F corresponding to 21 degree in Celsius. As the temperature increases, the dotted line shows an upward trend. Concretely, the dotted line shows the number of heat-related deaths in the temperature range that exceeds 21 °C: the higher the temperature, the higher the mortality rate.

What is intriguing is that the authors of the above study find that the number of heat-related deaths declined by almost 80% from the 1929–1959 period to the post-1959 period [1960–2004]. The mortality rate has changed from the dotted line during the 1929–1959 period to the dashed line during the 1960–2004 period. This shift is marked in the figure by a broad black arrow. The authors attribute this remarkable decline to the advances in air-conditioning during the latter half of the twentieth century (Barreca et al. 2016).

As far as the flu vaccines are concerned, there are quadrivalent vaccines and trivalent vaccines. The quadrivalent vaccines are the flu vaccines that protect against four different flu viruses: an influenza A (H1N1) virus, an influenza A (H3N2) virus, and two influenza B viruses. The trivalent vaccines are the vaccines that protect against three viruses. The quadrivalent vaccines explained by the CDC are Afluria Quadrivalent, Fluarix Quadrivalent, FluLaval Quadrivalent, and Fluzone Quadrivalent, Flucelvax Quadrivalent, and Flublok Quadrivalent (CDC 2020a).

For both malaria and the flu, the vaccines that are currently available give us some relief in the fight against these diseases. The large number of annual flu deaths in the US shown in Table 11.2 tells us, on the other hand, that the vaccine availability is not the final solution. Some vaccines may not be effective on some people. Many people do not or cannot get the vaccine shot. An effective treatment is as important as an effective vaccine.

11.5 Zoonoses for Farm Animals: Sleeping Sickness and Tsetse Flies

The concern on global warming-caused infectious diseases goes beyond the *anthroponoses*, the microbes that spread from human to human. The animal infectious diseases, also referred to as *zoonoses*, spread from animal to animal, sometimes wreaking havoc on a local/national economy by killing thousands and even millions of farm animals. Some animal diseases can be transmitted from animals to humans.

Because I want to put an emphasis on the human-to-human infectious diseases in this chapter and described previously animal diseases in the context of agriculture in Chapter 2, I will explain here briefly the nagana disease of African cattle, among numerous livestock diseases, *e.g.*, a bluetongue disease in sheep and goats, Australian cattle tick, and a foot-and-mouth disease (USAHA 2008).

The sleeping sickness, also commonly called nagana, is a chronic debilitating disease of cattle in Sub-Sahara. The primary disorder is anemia. The disease pathogen is carried by tsetse flies. Both tsetse flies and nagana are unique to Sub-Sahara. Unless treated, most cattle will eventually die. As such, the disease is a severe constraint to the livestock productions in Sub-Sahara.

The nagana is caused by one of the three types of *protozoan* parasites, i.e., single celled organisms, which belong to genus *trypanosoma*. The disease spreads from a cow to another through the bites of an infected tsetse fly (Ford and Katondo 1977).

There are several drugs available for treatment of the nagana infected animals, most commonly used of which are isometamidium chloride (trade name Samorin) and diaminazene aceturate (trade name Berenil) (USAHA 2008).

On another research track, scientists successfully sequenced the complete genomic code of the tsetse fly, which is seen as the first step toward controlling the distribution of tsetse flies in Sub-Sahara by modifying a part of the genome (Aksoy et al. 2014). This in turn offers another possibility to control the contagion of the nagana disease among Sub-Saharan cattle.

11.6   Pandemics and Novel Viruses in the Twenty-First Century

The challenges of the infectious diseases and lethal viruses reviewed in Sects. 11.2, 11.3, 11.4, and 11.5 have been recognized by the human society for a long time. With the dawn of the twenty-first century, the challenges may have entered a new more arduous phase. With the entry into the century, the global community was struck repeatedly by a series of novel viruses one after another, some of which have turned out to be highly contagious and at the same time very deadly: specifically, the

SARS (Severe Acute Respiratory Syndrome) virus in 2003, the Swine Flu (H1N1) in 2009, the MERS (Middle East Respiratory Syndrome) virus in 2012, and the novel coronavirus (COVID-19) in 2020.

Table 11.3 summarizes the pandemics of the twenty-first century and compares them against the deadliest pandemic of the twentieth century, the 1918–1920 pandemic commonly called the Spanish Flu. A pandemic is defined as "the worldwide spread of a new disease," which is declared by the World Health Organization (WHO). The SARS, the MERS, and the COVID-19, all of which were declared a pandemic, are all caused by different strains of the coronavirus, of which there are multiple: SARS-CoV, MERS-CoV, SARS-CoV-2. The coronavirus may have existed for millions of years on the Planet in bats and avian species. Many human coronaviruses have originated from bats. The coronaviruses are commonly called a bird flu or a bat flu. The term coronavirus comes from the shape of the virus which resembles the solar corona (CDC 2020c; JHU 2020).

The different types of the coronavirus have posed different levels of lethality. The fatality rate, specifically the case-fatality ratio, of the SARS-CoV was about 9% while that of the MERS-CoV was about 35%. These are extremely high mortality rates. The novel coronavirus is still unfolding, with the current mortality rate, as of 22 August 2020, of about 3.1% for the US, 13.8% for Italy, 1.9%% for India, and 1.7% for Japan, according to the Johns Hopkins University COVID-19 dashboard (JHU 2020). According to the US researchers, the mortality rate of COVID-19 is vastly overstated owing to many asymptomatic cases. A random sampling study indicates a mortality rate as low as 0.1% in the US. The asymptomatic cases are the COVID-19 infected persons but with no symptoms exhibited, which is common in COVID-19.

The novel coronavirus is on track to be the deadliest pandemic of the twenty-first century, with about 800,000 killed as of this writing. How does it compare with the Spanish Flu in 1918, the deadliest pandemic of the twentieth century? Like the Swine Flu of 2009, the Spanish Flu was caused by the H1N1 virus. The Center for Disease Control (CDC) estimates that at least 50 million people were killed by the 1918–1920 pandemic from about 500 million people infected by the virus. The 500 million people in 1918 accounted for about one-third of the world's population (Potter 2001; CDC 2020c).

Why have the pandemics of the twenty-first century resulted in far smaller numbers of human death than the Spanish Flu of the twentieth century? Noting that there is still no rigorous statistical study or consensus

Table 11.3 Novel Viruses of the twenty-first century vs. the Spanish Flu

Pandemic	Year	Virus	Origin	#of cases	#of Fatality	Mortality rate: case-fatality ratio
SARS (Severe Acute Respiratory Syndrome)	February 2003	SARS-associated coronavirus (SARS-CoV)	Asia	8,098	774	
H1N1, Swine Flu	Spring 2009	H1N1pdm09 virus	Mexico	60.8 million (US)	12,469 (US); 151,700–575,400 (World)	
MERS (Middle East Respiratory Syndrome)	2012	Middle East Respiratory Syndrome Coronavirus (MERS-CoV)	Saudi Arabia	2494	858	3 or 4 out of every 10
Novel Coronavirus (COVID-19)	2020	severe acute respiratory syndrome coronavirus 2 (SARS-CoV-2)	Wuhan, China	22.9 million (As of 22 August 2020)	799,245 (As of 22 August 2020)	13.8% (Italy); 3.1% (US) (JHU 2020)
The Deadliest Pandemic of the Twentieth Century						
Pandemic of 1918–1920: Spanish Flu	1918: the greatest medical holocaust in history	H1N1 virus	Not known: China, US	Infected 50% of the world's population; 25% a clinical infection	40–50 million (Potter 2001)	

on this question, I need to make it clear that it is a difficult task to measure correctly the infection rate as well as the number of fatalities from a pandemic. To give you an idea, the positivity/negativity tests may have been conducted neither randomly nor broadly across the society because, *among other reasons*, many infected people are asymptomatic. For another, the cause of death could be difficult to determine especially for a patient who long suffered from an existing disease. It is possible that many people who died of another disease during the Spanish Flu outbreak were wrongly identified to be a Spanish Flu casualty.

With these caveats in mind, I feel inclined to attribute the smaller number of fatalities of the twenty-first-century pandemics, at least up to this point in the century, to advances in medicine and medical systems (hospital capacities and treatments), advances in virology (tests and vaccines), a reduction in poverty rate, and an emergence of a mobile smartphone communication with which the pandemic information can be delivered directly to each person (Henderson et al. 2009).

11.7 FINAL WORDS ON THE CLIMATE CONNECTION

Although climate activists may be compelled to argue that the pandemics and novel viruses of the twenty-first century are being caused by the ongoing climate change of the Planet (Ban and Verkooijen 2020), there is no scientist who is able to associate the infectious diseases and viruses of this century that we witnessed with the ongoing climate change. The evidence and experiences point to a different conclusion. That is, advances in medical sciences, virology, a healthy economy, and a public health education are the best platform to fight against deadly viruses, considering that the current sciences cannot stop such infectious diseases from breaking out and the vaccine development after the breakout of a pandemic takes a long time, say, four years.

I am also inclined to suggest that climate activists should refrain from, considering the science and experiences, making an alarmist proclamation that the climate change in the twenty-first century will result in either far more frequent outbreaks of pandemics or far more lethal ones, but instead recognize the reality that there are other grave challenges faced by humanity other than climate change, of which the pandemic is one.

11.8   CHAPTER HIGHLIGHTS

- A pandemic is often the deadliest event that the human society experiences, even more fatal than great wars and severe hurricanes. It is estimated that the Spanish Flu in 1918 killed 50 million people worldwide and the novel coronavirus in 2020, occurring at the time of this writing, already killed about 800,000 people.
- Some climate activists may be quick to argue that the pandemics and novel viruses of the twenty-first century are being caused by the ongoing climate change of the Planet. But there is little scientific basis for the claim so far.
- This chapter provides a review of how the world community successfully eliminated the smallpox virus during the twentieth century.
- Among the anthroponoses, that is, human-to-human transmissible infectious diseases, two climate/weather associated diseases are reviewed: influenza and malaria. For both, vaccines are available.
- Of the zoonoses, that is, animal infectious diseases, we review the literature on African cattle sleeping sickness called nagana. Multiple vaccines are available.
- The pandemics that emerged since the dawn of the twenty-first century such as the SARS, Swine Flu, MERS, and COVID-19 are compared with the deadliest pandemic of the twentieth century, the Spanish Flu. The twenty-first-century pandemics appear to have resulted in far smaller numbers of fatalities.
- The malaria eradication in developed countries, a negative statistical relationship between flu-related deaths and flu-season average temperature, and the smallpox eradication experiences shed a ray of hope to the human society's fights against deadly viruses and pandemics.

REFERENCES

Aksoy, Serap, Geoffrey Attardo, et al. 2014. Genome Sequence of the Tsetse Fly (Glossina Morsitans): Vector of African Trypanosomiasis. *Science* 344 (6182): 380–386.

Associated Press (AP). 2008. *CDC to Destroy Oldest Smallpox Vaccine*. Published on March 3, 2008.

Ban, Ki-moon, and Patrick Verkooijen. 2020. *Will We Learn Lessons for Tackling Climate Change from Our Current Crisis?* On April 9, 2020. CNN.

Barreca, Alan, Karen Clay, Olivier Deschenes, Michael Greenstone, and Joseph S. Shapiro. 2016. Adapting to Climate Change: The Remarkable Decline in the US Temperature-Mortality Relationship over the Twentieth Century. *Journal of Political Economy* 124 (1): 105–59.

Centers for Disease Control and Prevention (CDC). 2020a. *Smallpox*. Washington, DC: The CDC.

Centers for Disease Control and Prevention (CDC). 2020b. *Influenza*. Washington, DC: The CDC.

Centers for Disease Control and Prevention (CDC). 2020c. *Past Pandemics*. Washington, DC: The CDC.

Ford, J., and K.M. Katondo. 1977. Maps of Tsetse Fly (Glossina) Distribution in Africa, 1973, According to Subgeneric Groups on a Scale of 1: 5000000. *The Bulletin of Animal Health and Production in Africa* 15: 187–93.

Henderson, D.A., Brooke Courtney, Thomas V. Inglesby, Eric Toner, and Jennifer B. Nuzzo. 2009. Public Health and Medical Responses to the 1957–58 Influenza Pandemic. *Biosecurity and Bioterrorism: Biodefense Strategy, Practice, and Science* 7 (3). http://doi.org/10.1089=bsp.2009.0729.

Iuliano A.D., K.M. Roguski, H.H. Chang, et al. 2018. Global Seasonal Influenza-Associated Mortality Collaborator Network. Estimates of Global Seasonal Influenza-Associated Respiratory Mortality: A Modelling Study. *Lancet* 391 (10127): 1285–1300.

Jenner, Edward. 1801. *On the Origin of the Vaccine Inoculation*. London, UK: D.N. Shury.

Johns Hopkins University (JHU). 2020. *COVID-19 Dash Board*. Baltimore, MD: JHU.

Lopez, Alan D., Colin D. Mathers, Majid Ezzati, Dean T. Jamison, and Christopher J.L. Murray. 2006. *Global Burden of Disease and Risk Factors*. New York, NY: World Bank and Oxford University Press.

Martens, P., R. Kovats, S. Nijhof, P. Devries, M. Livermore, D. Bradley, J. Cox, and A. McMichael. 1999. Climate Change and Future Populations at Risk of Malaria. *Global Environmental Change* 9 (October): S89–107.

Patz, J.A., A.K. Githeko, J.P. McCarty, S. Hussein, U. Confalonieri, and N. de Wet. 2003. Climate Change and Infectious Diseases. In *Climate Change and Human Health—Risks and Responses*. Geneva, CH: WHO.

Potter, C.W. 2001. A History of Influenza. *Journal of Applied Microbiology* 91 (4): 572–79.

Rogers, David J., and Sarah E. Randolph. 2000. The Global Spread of Malaria in a Future. *Warmer World. Science* 289 (5485): 1763–66.

Schwartz, M. (2001). The Life and Works of Louis Pasteur. *Journal of Applied Microbiology* 91 (4): 597–601.

United States Animal Health Association (USAHA). 2008. *Foreign Animal Diseases—The Gray Book*. MO: Committee on Foreign and Emerging Diseases of the USAHA.

World Health Organization (WHO). (2019). *World Malaria Report 2019*. Geneva, CH: WHO.

World Health Organization (WHO). 2020. *Malaria Vaccines*. Geneva, CH: WHO. Accessed from https://www.who.int/immunization/research/development/malaria/en/.

CHAPTER 12

Climate Negotiations: The Science of a Big Deal?

12.1 INTRODUCTION TO A GLOBAL DEAL ON CLIMATE CHANGE

When it comes to addressing global climate change, you will most certainly be quick to imagine a global grand policy in which every country participates and under which every carbon emitting activity is regulated for the sake of the Planet. Policy-makers and climate scientists have pursued such a big deal for over 30 years but without success. Every year, they gather in a famed city for negotiations during the two weeks right before Christmas holidays. Why couldn't they deliver good news the world is waiting for?

Many aspects of the global climate change make it reasonable for us to wish for a global cooperative policy. The greenhouse blanket is formed in the atmosphere and covers the entire globe. The carbon dioxide concentration of the Earth's atmosphere is quickly equalized across the globe. A shift in the climate system will affect all countries on the Planet. A single country's or even a single continent's efforts to cut greenhouse gases will be of no effect on the global climate system.

Considering all these, it may seem quite irrational to you that the world policy-makers failed to deliver an effective global treaty on climate change. The present author will explain in this chapter why and how such failures recurred by rational negotiators. To do this, we will together have to go through the climate policy negotiations since the 1980s.

© The Author(s), under exclusive license to Springer Nature
Switzerland AG 2021
S. N. Seo, *Climate Change and Economics*,
https://doi.org/10.1007/978-3-030-66680-4_12

12.2   THE FIRST IPCC REPORT ON CLIMATE CHANGE

The global warming problem for the Planet came to the recognition of perhaps only a handful of pioneer scientists before the 1980s. Notable is Svante Arrhenius who posited the causal relationship between the atmospheric temperature and carbon dioxide concentration as early as the late nineteenth century (Arrhenius 1889). Another are Revelle and Suess who formulated a grand model of the carbon exchange between the atmosphere and the world oceans (Revelle and Suess 1957).

The global warming became a recognized world-wide policy problem through the 1980s through UN-organized international conferences, which culminated with the establishment of the Intergovernmental Panel on Climate Change (IPCC). Soon afterward, the IPCC issued its first set of reports in 1990 (IPCC 1990a). The decade witnessed multiple international climate conferences and the publications of the earliest scientific papers which attempted to predict the degree of climate change far in the future, say, centuries later.

The first assessment report (FAR) of the IPCC, composed of three volumes, summarized the cutting-edge sciences of global warming up to that point, including the future predictions of climate change, which have turned out to lay the foundation for all the succeeding publications by the IPCC. The IPCC assessment report is published every 5 year or so and the sixth assessment report is under preparation at the moment.

Specifically, the first assessment report contained the expositions of the now common terms and approaches, which at the time alien and unheard of to everyone, such as greenhouse gases, greenhouse effect, radiative forcing, global warming potential, carbon cycle, historical climate change, predictions of future climate change, and future scenarios.

The first assessment report had three volumes, with the first volume devoted to scientific assessments of climate change, the second volume devoted to impact assessments, and the third volume devoted to response strategies. Each volume was written by the respective working group (WG): WG 1, WG 2, and WG 3. This initial arrangement of the report structure has remained unchanged through the sixth assessment report that is under preparation.

Another format-setting feature of the FAR which has continued in its ensuing publications was its policy focus. In addition to the full report, the IPCC presented shorter versions directed to policy-makers. One was the technical report for each volume. Another was the summary for policy-makers (SPM).

The most shocking, as well as, contentious of all the contents in the IPCC's first assessment report was the predictions of future global temperature increases through year 2100. The figure showed that the best estimate of the IPCC is around 4.3 °C increase in global average temperature by 2100. The high-end estimate and the low-end estimate amounted to 6.3 °C increase and 3 °C increase (IPCC 1990a).

Another critical and enduring contribution of the IPCC first assessment report was the identification of sea level rise and agriculture as the most vulnerable sectors of the society under climate change. The volumes 1 and 2 of the FAR both highlighted the two areas of concern. The section on the vulnerability of agriculture and forestry headed the impact assessment volume by the WG 2 (IPCC 1990b; Seo 2017a).

12.3 THE ESTABLISHMENT OF THE UNFCCC INSTITUTION

The establishment of the IPCC and the publication of the first assessment report had laid, in retrospect, a solid platform from which policy discussions among the nations of the Planet could begin. The momentum was captured by the United Nations Conference on Environment and Development in Rio de Janeiro in summer 1992 through an agreement among the UN members to create a UN-led climate change convention.

The Rio Earth Summit, as the UN conference in 1992 is commonly referred to as, created ultimately the foundational institution through which climate change challenges can be addressed at a global level: The United Nations Framework Convention on Climate Change (UNFCCC) (1992). As of 2020, it has 197 parties to the convention, a nearly universal membership of the world nations.

Of the core accomplishments, the international treaty, *i.e.*, the treaty of the UNFCCC, declared, nearly 30 years ago, the objective of the convention unequivocally as follows, which is agreed upon by all the member nations (UNFCCC 1992):

The ultimate objective of this Convention ......is to achieve, in accordance with the relevant provisions of the Convention, stabilization of greenhouse gas concentrations in the atmosphere at a level that would prevent dangerous anthropogenic interference with the climate system.

Article 2 of the UNFCCC treaty, cited above, still provides the solid benchmark from which all policy discussions and negotiations of the world nations are conducted. I presume that readers are at this point of the book comfortable with the term greenhouse gas concentrations in the above article, so I do not need to explain here. All the efforts of the globe when it comes to climate negotiations are interpreted by most people as the endeavors to "prevent dangerous anthropogenic interference with the climate system" owing to the Article 2 declaration.

The primary responsibility of the UNFCCC is to organize international conferences during which policy responses and progresses of the global community can be discussed and negotiated. Of the variety of conferences and meetings convened by the convention, primary ones are the Conference of the Parties, which is referred to as the COP in short. The COP is normally held once a year and during the two weeks before Christmas holidays. The most recent COP was held in Madrid, Spain, which was COP 25.

From the inception of the UNFCCC, a lingering concern was identified and then marked unmistakably by the signatories of the treaty. That concern was how the world community should divide the burden of cutting greenhouse gas emissions among the nations. Considering differences of economy, politics, history, geography, and culture across the world nations, it was recognized by the signatories to be a tough one that cannot be addressed handily. Even so, the first negotiators in 1992 may have not seen a truly thorny nature of the concern as regards the global action, which would eventually unfold over the next three decades before our very eyes (Seo 2017b).

The thorny issue was expressed in the UNFCCC treaty as the first principle of the treaty among multiple such principles, which was expressed succinctly as "common but differentiated responsibilities" among the member nations. Concretely, the first principle is stated as follows (UNFCCC 1992):

> The Parties should protect the climate system for the benefit of present and future generations of humankind, on the basis of equity and in accordance with their common but differentiated responsibilities and respective capabilities. Accordingly, the developed country Parties should take the lead in combating climate change and the adverse effects thereof.

Of the 25 COPs that ensued, two of them resulted in a global agreement and a Protocol: the Paris Agreement in 2015 (COP 21) and the Kyoto Protocol in 1997 (COP 3). The Kyoto Protocol was signed as an international treaty, meaning that it has the legally binding force. The Paris Agreement was a non-binding voluntary agreement, as such, no approval from the congress of the individual party of the convention was required.

In the following sections, I will explain major outcomes from the two international agreements/treaties, which will reveal to you how the aforementioned objective and the principle of the UNFCCC have materialized in the global negotiations. Further, the review will reveal the core obstacles that have derailed the negotiations for a truly global and effective climate deal.

12.4   Kyoto Protocol: The Emergence of Economic Problems

The UNFCCC entered into force in 1994 and the first COP organized by the organization was held in Berlin in 1995. Spurred by the momentum created by the establishments of the IPCC and the UNFCCC, the world community achieved the biggest milestone in just two years in the global warming policy negotiations. At the COP 3 held in December 1997, the Kyoto Protocol was signed by the parties of the conference (UNFCCC 1997).

The signing of the Kyoto Protocol meant the emergence of the economics of global warming for the first time (Nordhaus 2001). Put differently, for the first time in global warming debates, each country started to assess its own economics of policy implementations encapsulated in the Kyoto Protocol. Signed as an international treaty that went through the congressional approval of each signatory, it dawned on each and every party of the conference that the negotiated outcomes might end up requiring a big sacrifice of the country while other countries might end up sacrificing little.

What was in the Kyoto Protocol which sparked such a sharp reaction from the individual countries? At the core of the Protocol lie the overarching goal which must be enforced with legal force: the reduction of carbon dioxide equivalent emissions by at least 5% below the 1990 emission level collectively. This had quickly become the focal point of intense debates among the nations and concerned people. Citing directly Article 3 of the Protocol (UNFCCC 1997):

The Parties shall, individually or jointly, ensure that their aggregate anthro-pogenic carbon dioxide equivalent emissions of the greenhouse gases......do not exceed their assigned amounts, ..., with a view to reducing their overall emissions of such gases by at least 5 per cent below 1990 levels in the commitment period 2008 to 2012.

The economic deliberations of the individual parties on what Article 3 contains arrived at two primary misgivings, which were well elucidated by the leading economists' works. The first was whether the Kyoto Protocol's approach of capping the global emissions at a certain level or year would be a hugely more expensive to the globe than the least cost policy, that is, a globally efficient global warming policy (Manne and Richels 1999; Nordhaus and Boyer 1999).

The second reservation of the signatories of the Protocol was whether the Kyoto approach would result in calling for a gargantuan sacrifice of some countries while other countries are exempted from the economic burden of taming the global climate system. In particular, the Kyoto approach exempted developing countries, including China and India, from any mitigation burden.

The paths of the parties of the conference started to diverge from that point on, that is, the signing of the Protocol. Looking back, three coali-tions may have emerged from the economic deliberations of the Kyoto Protocol. The first coalition was the European Union which proceeded with full force to implement the Kyoto agreement, which was material-ized later through the EU Emission Trading Scheme (ETS) (EC 2016). The ETS was implemented in the EU throughout the first phase of the Kyoto Protocol, from 2008 to 2012, and continued beyond.

The second coalition was represented by the US. The US Congress did not ratify the Kyoto Protocol and the country eventually withdrew from the Protocol. The rationale for the withdrawal decision was that the treaty imposes too high cost on the fast-growing economy of the US while it places only minor cost or even exempts other major fast-growing economies from the economic burden of a global climate policy, specif-ically, China and India. At the end of the first implementation period of the Kyoto Protocol, other countries followed the US lead and withdrew from the Kyoto agreement such as Japan, Australia, Canada, and Russia.

The third coalition may have been the group of countries that were exempted from the burden of cutting greenhouse gases at the Kyoto conference, including China and India. The third group countries were not included in the Annex B of the Protocol which specified the emission reduction commitments by the individual countries.

The Kyoto Protocol entered into force in 2005 with the ratification of Russia, without the US participation. This meant that a dominant fraction of the global emissions of carbon dioxide equivalents, released from China, India, and US, was not part of the Kyoto Protocol's policy implementation, making it of little effect when it comes to the goal of a global climate stabilization.

The divergence emerged in the first phase also signaled that the negotiations for the second phase of the Kyoto Protocol in which all member countries are included in the sharing of the global mitigation burden would be rocky, to say the least. A grand mobilization to reach such a comprehensive global deal took place in Copenhagen Conference in Denmark in 2009. That year coincided with the first year of the Obama Presidency, ending eight years of the Bush Presidency which decided to withdraw from the Kyoto Protocol, which raised the expectation through the roof.

Notwithstanding the favorable political winds and the high expectations set by the global media and negotiators, the Conference became "disarrayed" toward the tail end of it and "in shambles" in the end, setting the world media ablaze. The Conference could not overcome the disagreements among the nations on whether or not the Kyoto Protocol should be expanded to a global framework (UNFCCC 2009).

In retrospect, by the time of Copenhagen Conference in 2009, the parties of the conference had gotten substantially sophisticated about the economic calculus of the international climate treaty, both individually and collectively (Nordhaus 2010). The negotiators representing their respective countries had become cunning as a fox in protecting their nations' interests, that is, in just 12 years from the Kyoto Protocol's goodwill approach.

12.5 Paris Agreement: A Non-binding Big Deal?

The dramatic failure to reach a comprehensive deal for an expanded Kyoto Protocol at Copenhagen disappointed all the negotiators and concerned citizens of the globe. But, there was a silver lining. With Barak Obama assuming the Presidency of the US in 2009 and Xi Jinping the Presidency of China in 2012, both men of the two most powerful nations on the Planet signaled that they will support a global climate policy program.

The future of a global climate policy intervention seemed to be bright once again.

The renewed resolve of the global community was materialized into the Durban Platform for Enhanced Action at the Durban Conference in South Africa in 2011 (UNFCCC 2011). Through the Platform, nations agreed to launch a process to reach a truly global-scale deal by the end of 2015 that binds all countries legally to the mitigation requirements of greenhouse gases. The 2015 conference was then assigned to Paris, France.

The key lesson from the Copenhagen disappointment to the global negotiators was that the disagreements between developed nations and developing nations must be narrowed beforehand to have any shot at the kind of treaty promulgated in Durban. In particular, the developing nations including China demanded that the principle of "common but differentiated responsibilities" should be adhered to in any Paris deal, which stood like an elephant in the room.

Another sticking point espoused by the developing nations that had been set aside from the global mitigation burdens up to that point was that the rich country parties should assist financially the poor country parties as a precondition for their participation into the Paris deal. They argued that such monetary transfers would help the poor countries take expensive mitigation steps and adopt necessary adaptation measures. Further, their demand was linked to the emerging financial mechanism of the UNFCCC, that is, the Green Climate Fund (GCF). The GCF will be explained in the next section.

How did the Paris Conference turn out? Did the parties make a grand compromise for the sake of the Planet? In what ways the developed nations accommodated the tall demands by the developing nations? At the time of this writing nearly five years after it was concluded, I can say that the Paris Conference produced major policy outcomes and still gives us important lessons on a global policy-making process.

First and foremost, the Paris Agreement was achieved at the end of the Conference and signed by the 196 nations. Upon signing the Agreement, it was widely hailed, by the global media and world leaders, as a historic agreement, entering a new era of global cooperation, a turning point in climate fight, and even the greatest diplomatic success (UNFCCC 2015).

A more reasoned analysis of the document of the Paris Agreement, however, revealed that the nations might have chosen only a temporary truce that would be broken in any minute (Seo 2017a, b). In fact, it was clear to any long-time observer of the UN-led negotiations that the arrangements between the developing country parties and the developed country parties in the Agreement would not be sustainable for long.

Unable to formulate a forceful framework that matches the Kyoto Protocol, negotiators devised a framework named the "Intended Nationally Determined Contributions (INDC)." This meant that the promised actions of the individual nations inscribed in the Paris Agreement would be determined by the nations themselves, that is, not by a globally determined policy goal. Further, the INDCs declared by individual nations were voluntary, needing neither monitoring by an international body nor penalties for not complying with them.

As such, the INDCs submitted by the individual parties varied widely in ambitions and policy targets. Most notably, many influential players such as China, India, and South Africa did not commit to emission reduction targets. For another, many developing nations proposed emission reduction targets, but on the pre-conditions set by the nations themselves. Figure 12.1 summarizes the INDCs pledged by the major negotiating parties to the Paris Agreement.

Let me explain the INDCs of the parties by three groups. The first group is the US and the EU. The US proposal is to cut the emissions from carbon dioxide equivalents by 26–28% by 2025. The EU's proposal is to cut the emissions thereof by 40% by 2030. Note in the figure that

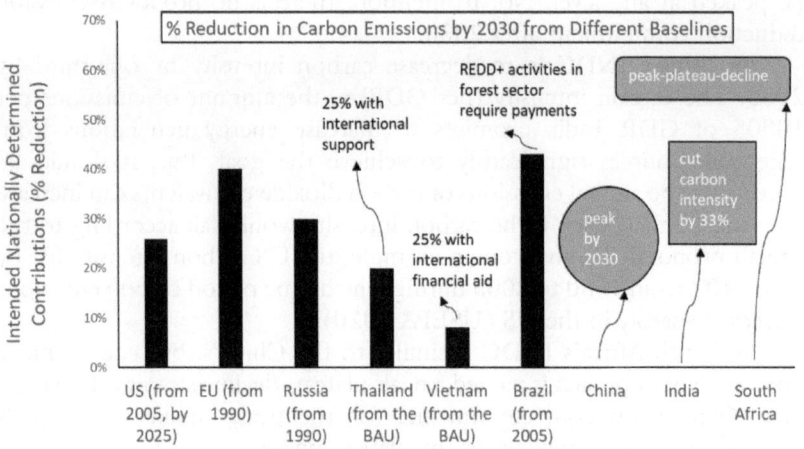

Fig. 12.1 Intended Nationally Determined Contributions to the Paris Agreement

the two regions use different baselines: the 2005 emissions level for the former and the 1990 emissions level for the latter.

The second group is the group of developing countries: Thailand, Vietnam, Brazil, and Russia. As described in the figure, the INDCs submitted by the group 2 countries are attached with important qualifications or conditions. Concretely, the Vietnam's INDC is an 8% reduction unconditionally, but increases to a 25% reduction on the condition of international financial aids through, say, the GCF. Similarly, the Thailand's INDC is a 20% reduction unconditionally but increases to a 25% reduction on the condition of international technical and financial supports. The Brazil's INDC is a 43% reduction in the emissions by 2030 unconditionally, which nonetheless comes with an important qualification. Brazil requests that the forest sector carbon credits generated from the Brazilian forests should be given payments and the emission reduction credits generated from Brazil cannot be used by other countries for their INDC commitments (UNFCCC 2016).

The third group of countries in the figure is the group of three fast-growing economies: China, India, and South Africa. All three countries did not propose concrete emission reduction commitments. The China's INDC is simply to peak the emissions of carbon dioxide equivalents by 2030. This means that China should be exempted from any emission reduction responsibility up to the year 2030 while the 2030 emission can be peaked at any level. Not to mention, there is no proposed emission reduction commitment after 2030.

The India's INDC is to decrease carbon intensity by one-third by 2030. The carbon intensity (per GDP) is the amount of emissions per 1000$ of GDP. India promises to increase energy generations from renewable sources significantly to achieve the goal. But, it should be noted that the annual emissions of carbon dioxide equivalents can increase very substantially even if the carbon intensity would fall according to the Indian proposal. To give you an example, the US carbon intensity fell by about 40% from 1980 to 2005 during which time period carbon emissions increased sharply in the US (USEPA 2020).

The South Africa's INDC is similar to the China's, but is even more obscure. The approach is named a peak-plateau-decline strategy. In particular, there is no concrete timeline for the peak, unlike the China's proposal, neither a timeline for the plateau phase.

As the initial jubilation on the outcomes of the Paris Conference faded, concerned scientists and citizens began to ask the inevitable question:

Does the INDCs summed across all countries amount to anything, that is, any meaningful emission reduction in the control of the climate warming? The IPCC and other environmental organizations of the United Nations, among others, took on the question. The answer was the sum of the INDCs, even if all commitments of the countries are faithfully executed, may not mean much in terms of the degree of reduction in Planetary warming. The two reports are referred to as the emissions gap report for the UNEP publication and the 1.5 °C report for the IPCC publication (UNEP 2017; IPCC 2018).

The latter report should grasp your attention: Why is it called a 1.5 °C report? This is a sharp question that pertains to the ultimate goal of the Paris Agreement. To put the question differently: What is the ultimate goal envisioned by the Paris Agreement? That is the ceiling of 1.5 °C in the control of Planet warming.

The Paris negotiators adopted the temperature ceiling, which was first introduced at the Cancun Conference in 2010, the next year after the Copenhagen disappointment, as the ultimate objective of a global climate policy (UNFCCC 2011). For now, it suffices for me to say only that whether the temperature ceiling is an appropriate climate policy target is being debated by researchers. To give you an idea, it would be reasonable to ask whether a 1.5 °C temperature ceiling would be very different from a 3 °C temperature ceiling in terms of, inter alia, the Planet's climate stability and the economic impacts of climate change.

Another contentious point, which emerged from the Paris Agreement, in subsequent UN conferences was a matter of monitoring and inspection. Simply put, for any effective global policy, an international body of inspectors should be allowed to monitor and verify the emission reduction claims made by each party of the conference. However, the UN conferences could not make that a reality owing to the objections from many countries on the grounds of national sovereignty. This means that the Paris Agreement is neither effective nor verifiable to be a trustworthy mechanism to ensure the Planet's climate wellness.

12.6 THE GREEN CLIMATE FUND

I interspersed the term "Green Climate Fund" at several places in this chapter without a formal explanation of it. Emerging from the final days of the Copenhagen Conference, it has surged in the past few years to the forefront of the climate policy negotiations, especially after the Paris

Conference (Seo 2019a). The reason for the surge and enthusiasm for the GCF is not difficult to fathom. Given the failures and disappointments of the global negotiations that the parties have witnessed for three decades, the GCF was by contrast a real thing, that is, actual dollar transfers from one country to another.

It started as a surprise announcement by the US during the final days of the Conference and came to be written into the Copenhagen Accord as the Copenhagen Green Climate Fund. It was originally a pledge to generate US$ 100 billion annually to assist poor countries to adapt to numerous challenges from climate change (UNFCCC 2009). In the next several years, the GCF was formally established under the UNFCCC: The GCF governing instrument was approved; the secretariat was approved and built in Songdo City, South Korea; the governing board composed of 24 members was selected; the pledges were collected; disbursements of the GCF fund began at the end of 2015 just before the Paris Agreement (Seo 2019a).

As of March 2020, 129 projects selected internationally by the GCF Board either received or are scheduled to receive a grant or a loan, mostly a grant, from the organization. The total funding committed by the GCF up to the point to the list of selected projects is US$ 5.6 billion. The projects are selected from the pool of proposals submitted by the developing country applicants from Africa, Latin America, South Asia, Pacific Islands, Central Asia, and Eastern Europe (GCF 2020b).

In Tables 12.1 and 12.2, I summarize the GCF-funded projects proposed by African countries. For the sake of a lucid explanation, I chose the projects in the "results areas" of energy generation and access as well as of food security. The results areas are defined by the GCF and each application should specify which areas the proposal addresses. As you can verify, a vast majority of the energy projects in Africa is a solar project or another renewable energy project. The funding to thee projects is in the range of US$39 million at the low end to 1 billion at the high end.

On the other hand, the food security projects shown in Table 12.2 are varied across the countries, encompassing climate information service, women, smallholders, crops, rangelands, orchards, irrigation, rivers, ecosystems, and agriculture funds. Many of the food security projects in the table aim for a resilient system to climate risk. Many of them are also directed to the most vulnerable regions and systems in the applicant country for which the GCF funds are sought.

Table 12.1   GCF-funded projects on energy generation and access in Africa (As of March 2020)

Countries of the proposal	Project title	GCF results area	Funding size (m. US$)
Burkina Faso	Yeleen Rural Electrification Project in Burkina Faso	Energy	60.1
DR Congo	DRC Green Mini-Grid Program	Energy	89.0
Mali	Mali solar rural electrification project	Energy	38.7
Nigeria	Nigeria Solar IPP Support Program	Energy	467.0
South Africa	Embedded Generation Investment Programme (EGIP)	Energy	537.0
Benin, Burkina Faso, Guinea-Bissau, Mali, Niger, Togo	BOAD Climate Finance Facility to Scale Up Solar Energy Investments in Francophone West Africa LDCs	Energy	138.0
Benin, Kenya, Namibia, Nigeria, Tanzania	Universal Green Energy Access Programme	Energy	301.6
Mauritius	Accelerating the Transformational Shift to a Low-Carbon Economy in the Republic of Mauritius	Energy	191.4
Egypt	Egypt Renewable Energy Financing Framework	Energy	1000.0
Rwanda	Strengthening climate resilience of rural communities in Northern Rwanda (SCRNRP)	Energy	33.2
Zambia	Zambia Renewable Energy Financing Framework	Energy	154.0

Table 12.2 GCF-funded projects on food security in Africa (As of March 2020)

Countries of the proposal	Project title	GCF results area	Funding size (m. US$)
Morocco	Development of Argan orchards in Degraded Environment—DARED	Food security	49.2
Zimbabwe	Integrated Climate Risk Management for Food Security and Livelihoods in Zimbabwe focusing on Masvingo and Rushinga Districts	Food security	10.0
Mozambique	Climate-resilient food security for women and men smallholders in Mozambique through integrated risk management	Food security	10.0
Gambia	Large-scale Ecosystem-based Adaptation in the Gambia River Basin: developing a climate resilient, natural resource-based economy	Food security	25.5
Mali	Africa Hydromet Program—Strengthening Climate Resilience in Sub-Saharan Africa: Mali Country Project	Food security	27.3
Namibia	Climate Resilient Agriculture in three of the Vulnerable Extreme northern crop-growing regions (CRAVE)	Food security	10.0
Namibia	Empower to Adapt: Creating Climate-Change Resilient Livelihoods through Community-Based Natural Resource Management in Namibia	Food security	10.0

(continued)

Table 12.2 (continued)

Countries of the proposal	Project title	GCF results area	Funding size (m. US$)
Morocco	Irrigation development and adaptation of irrigated agriculture to climate change in semi-arid Morocco	Food security	93.3
Senegal	Building the Climate Resilience of Food-insecure Smallholder Farmers through Integrated Management of Climate Risks (the R4 Rural Resilience Initiative)	Food security	10.0
Ethiopia	Responding to the Increasing Risk of Drought: Building Gender-responsive Resilience of the Most Vulnerable Communities	Food security	50.0
Zambia	Strengthening climate resilience of agricultural livelihoods in Agro-Ecological Regions I and II in Zambia	Food security	137.3
Uganda, Ghana, Nigeria	Acumen Resilient Agriculture Fund (ARAF)	Food security	56.0
Namibia	Improving rangeland and ecosystem management practices of smallholder farmers under conditions of climate change in Sesfontein, Fransfontein, and Warmquelle areas of the Republic of Namibia	Food security	10.0
Zimbabwe	Building Climate Resilience of Vulnerable Agricultural Livelihoods in Southern Zimbabwe	Food security	47.8

Despite the initially declared size of the Fund, the pledges made by the rich country parties have fallen far short of the ambition. As of October 2020, the confirmed contributions during the initial resource mobilization period totaled around US$ 8.2 billion, committed mostly in 2014 and in grants. With the withdrawal of the US from the Paris Agreement, the GCF resource mobilization also suffered a blow (Seo 2019b). The first replenishment of the GCF, named the GCF-1, began for the 2020–2023 period, which raised US$ 5.3 billion in confirmed pledges from mostly EU countries and Japan. It is not clear what fraction of the pledges is made in grants (GCF 2020a).

As we saw in the Paris Agreement's INDCs described above, the developing country parties started to see a grant from the GCF as a precondition for any climate change mitigation and adaptation action to be undertaken in their countries. In Fig. 12.1, you can verify this from Thailand's, Vietnam's, and Brazil's conditional commitments. From this viewpoint, the GCF may have created a wrong incentive to the poor countries by encouraging them, albeit unintentionally, not to do anything unless there is a dollar transmitted from the GCF account to a specific project's account (Seo 2015, 2019a).

Further, a closer examination of the GCF-funded projects one by one reveals that the GCF grant/loan is likely to make the funding recipients more vulnerable to climate change rather than to make them less vulnerable. Why would it be so? This is because the grants are allocated in most occasions to the most vulnerable sectors, regions, and populations of the recipient countries. Therefore, the GCF grants would consequently encourage the recipients to stick to the currently vulnerable systems and the regions.

From a larger point of view, this brings us to the question of the soundness of the GCF investment decisions. The Fund publishes the set of investment criteria used for their decisions which was formulated on the basis of the Governing Instrument approved at Durban Conference in 2011. Based on the "scores" given to the criteria, the GCF Board decides on a unanimous decision of the 24 board members whether it will approve or not the project under consideration. The Board members are selected equally from the developing country parties and the developed country parties of the UNFCCC, that is, 12 from the former and 12 from the latter (GCF 2011).

The investment criteria, however, do not give you a clarification or confidence because there are 7 criteria and roughly three dozen sub-criteria considered by the Board and further the weights assigned to the criteria are not specified nor revealed by the Board. Further, the set of investment criteria does include the criteria that are little directly related to climate change, climate mitigation, or climate adaptation: for example, the needs of a country, country ownership, sustainable development goals, among other things (Seo 2019a).

At this point in time in the development of the GCF, we should ask whether the GCF funding activities could minimize, as hoped for, the disagreements between the developing country parties and the developed country parties for the sake of a global climate treaty, the chief motivation for the inception of the GCF. The GCF outcomes, on the contrary, may widen the disparities between the two if it would turn out to distort efficient adaptation decisions of individuals in the recipient countries (Seo 2015).

12.7 OVERCOME CHALLENGES OR SEEK AN ALTERNATIVE PATH?

The retrospection thus far of the global negotiations on climate change for the past three decades reveals that the global community has gradually come to grips with the harsh reality of climate negotiations: There exists a big gap in the viewpoints between the developed country parties and the developing country parties on how the human society should tackle the problems of climate change. Further, even within each cohort of the two parties, there often exists a sizeable gap among the nations in their policy priorities and preferred actions.

The disparities put the global negotiations in bind by pushing individual parties to take non-cooperative actions. After the US withdrawal from the Paris Agreement, which was followed by the changes in the policy positions of Australia and Brazil, the global level negotiations for an international climate treaty are virtually stalled (White House 2017). At the COP 25 in Madrid in December 2019, China, India, Brazil, and South Africa issued a joint statement denouncing "imbalances in negotiations," specifically, a lack of financing commitments by rich nations (SCMP 2020).

At this moment in time, I feel it opportune to ask whether the global negotiations should continue with our eyes fixated on arriving at a grand global climate treaty in which all parties are embraced and all parties are enthusiastic and willing to bear legal responsibilities of cutting Planet-heating gases. If not, you may wonder, what alternative ways are there anyway? We can frame the question more formally as follows: Is there an alternative path forward for the Planet's community which can render as effective a global response to the challenges of climate change as any future global climate treaty?

The present author has reflected upon this question for a "long" time and recently presented a novel framework rooted on individual decision-makers and their behavioral changes (Seo 2020). This novel framework sheds light on four economic decisions: microbehavioral efficient decisions, forward-looking decisions, prescient decisions, and technological innovations. If you are an astute reader, you will see that the four decisions are well accounted for in this book. For the exposition of the technological innovations, you can refer to the chapters on greenhouse technologies and energy revolutions (Chapters 7 and 8). For the exposition of the microbehavioral efficient decisions, you may refer to the chapters on Sub-Saharan animals, the Amazon forests, and the Indian monsoon (Chapters 3–5). For the exposition of the forward-looking decisions and prescient decisions, you may go back to review the chapters on, forests, grasslands, and tropical cyclones (Chapters 3, 5, and 6).

12.8   Chapter Highlights

- This chapter provides a wide-ranging review of the global negotiations on climate change, with the key outcomes from the United Nations Framework Convention on Climate Change, the Kyoto Protocol, the Copenhagen Accord, the Paris Agreement, and the Green Climate Fund explained.
- The biggest hurdle for a global climate treaty continues to be the strong disagreements between developed country parties and developing country parties on who should bear the costly burden of cutting greenhouse gas emissions for the Planet.
- The Green Climate Fund has surged as a primary mechanism of the UNFCCC to bridge the two parties through a batch of monetary transfers now and then from the funds donated by the rich countries to a bundle of the climate projects proposed by the developing country parties.
- With the withdrawal of the US from the Paris Agreement, which was followed by a series of policy changes in Australia, Brazil, China, and others, the gap between the two negotiation cohorts has widened.
- Considering the repeated failures in search for a global climate treaty it would be opportune to explore a novel alternative path rooted on

individual decision-makers and their behavioral changes in striving to adapt to and utilize the climate-related resources.

References

Arrhenius, S.A. 1889. Ü¨ber die Dissociationsw¨arme und den Einfluß der Temperatur auf den Dissociationsgrad der Elektrolyte. *Zeitschrift für Physikalische Chemie* 4, 96–116.

European Commission (EC). 2016. *EU ETS Handbook*. Brussels: European Commission.

Green Climate Fund (GCF). 2011. *Governing Instrument for the Green Climate Fund*. Songdo, ROK: GCF.

Green Climate Fund (GCF). 2020a. *Status of Pledges and Contributions Made to the Green Climate Fund, Status Date: 31 July 2020*. Songdo, ROK: GCF. Accessed from https://www.greenclimate.fund/document/status-pledges-all-cycles.

Green Climate Fund (GCF). 2020b. *Project Portfolio*. Songdo, ROK: GCF. Accessed from https://www.greenclimate.fund/projects.

Intergovernmental Panel on Climate Change (IPCC). 1990a. *Climate Change: The IPCC Scientific Assessment*. Cambridge, UK: Cambridge University Press.

Intergovernmental Panel on Climate Change (IPCC). 1990b. *Climate Change: The IPCC Impacts Assessment*. Canberra, AU: Australian Government Publishing Service.

Intergovernmental Panel on Climate Change (IPCC). 2018. *Special Report on Global Warming of 1.5 °C*. Cambridge, UK: Cambridge University Press.

Manne, Alan S., and Richard G. Richels. 1999. The Kyoto Protocol: A Cost-effective Strategy for Meeting Environmental Objectives? *The Energy Journal* 20 (Special Issue): 1–23.

Nordhaus, William. 2001. Global Warming Economics. *Science* 294: 1283–84.

Nordhaus, William. 2010. Economic Aspects of Global Warming in a Post-Copenhagen Environment. *Proceedings of the National Academy of Sciences of the United States* 107 (26): 11721–26.

Nordhaus, William D., and Joseph G. Boyer. 1999. Requiem for Kyoto: An Economic Analysis of the Kyoto Protocol. *Energy Journal* 20 (Special Issue): 93–130.

Revelle, R., and H.E. Suess. 1957. Carbon Dioxide Exchange Between Atmosphere and Oceanand the Question of an Increase of Atmospheric CO2 During the Past Decades. *Tellus* 9, 18–27.

Seo, S. Niggol. 2015. Helping Low-latitude Poor Countries with Climate Change. *Regulation*. Winter 2015–2016: 6–8. Cato Institute, Washington, DC.

Seo, S.Niggol. 2017a. *The Behavioral Economics of Climate Change: Adaptation Behaviors, Global Public Goods, Breakthrough Technologies, and Policy-making*. Amsterdam, NL: Academic Press.

Seo, S.Niggol. 2017b. Beyond the Paris Agreement: Climate Change Policy Negotiations and Future Directions. *Regional Science Policy & Practice* 9 (2): 121–40. https://doi.org/10.1111/rsp3.12090.

Seo, S.Niggol. 2019a. *The Economics of Global Allocations of the Green Climate Fund: An Assessment from Four Scientific Traditions of Modeling Adaptation Strategies*. Cham, CH: Springer Nature.

Seo, S.Niggol. 2019b. Economic Questions on Global Warming During the Trump Years. *Journal of Public Affairs* 19 (1): e1914. https://doi.org/10.1002/pa.1914.

Seo, S.Niggol. 2020. *The Economics of Globally Shared and Public Goods*. Amsterdam, NL: Academic Press.

South China Morning Post (SCMP). 2020. *COP25 Summit: Expected to be Cooling Influence at UN Climate Conference, China Instead Lets Brazil Heat Up*. Hongkong: SCMP. Accessed from https://www.scmp.com/news/china/diplomacy/article/3041901/cop25-summit-expected-be-cooling-inf luence-un-climate.

United Nations Environment Programme (UNEP). 2017. *The Emissions Gap Report 2017: A UN Environment Synthesis Report*. Nairobi, KE: UNEP.

United Nations Framework Convention on Climate Change (UNFCCC). 1992. *United Nations Framework Convention on Climate Change*. New York, NY: UNFCCC.

United Nations Framework Convention on Climate Change (UNFCCC). 1997. *Kyoto Protocol to the United Nations Framework Convention on Climate Change*. New York, NY: UNFCCC.

United Nations Framework Convention on Climate Change (UNFCCC). 2009. Copenhagen Accord. Geneva: UNFCCC.

United Nations Framework Convention on Climate Change (UNFCCC). 2011. The Durban platform for enhanced action. New York: UNFCCC.

United Nations Framework Convention on Climate Change (UNFCCC). 2015. The Paris Agreement. Conference of the Parties (COP) 21. New York, NY: UNFCCC.

United Nations Framework Convention on Climate Change (UNFCCC). 2016. *Key Decisions Relevant for Reducing Emissions from Deforestation and Forest Degradation in Developing Countries (REDD +)*. New York, NY: UNFCCC.

United States Environmental Protection Agency (USEPA). 2020. *Inventory of US Greenhouse Gas Emissions and Sinks: 1990–2018*. Washington, DC: USEPA.

White House. 2017. *Statement by President Trump on the Paris Climate Accord*. Washington, DC: White House.

CHAPTER 13

Climate Change and Economics with Young Enthusiasts: Inter-generational Gaps and Burden Sharing

13.1 What Does It All Mean?

Together, we have walked through in this book a whole lot of not only physical terrains but also intellectual realms as they concern the Planet's climate challenges. As a matter of fact, the book was first conceived as a small book which explains lucidly all the complexities of climate change and economics and simultaneously is easy to read. Now is the time for us to celebrate a grand finale!

I began this book with an invitation to you for a pleasant stroll. We have seen together, to refresh your memory, animals of Sub-Sahara, Amazon rainforests, an Indian monsoon, semi-dwarf wheat, Bay of Bengal cyclones, North Atlantic hurricanes, world oceans, Pampas grasslands of South America, grand rivers of Asia, fossil fuels-based and alternative energy technologies, the massive Three Gorges dam in China, solar reflectors high up in the stratosphere, polar bears in the Arctic Ocean, coral reefs in the Coral Triangle, Antarctica's ice sheets, fishermen and fisheries, fatal invisible viruses, infectious diseases, and negotiators' roundtables in Paris.

The present author concluded each chapter with a closure within the purview of the main topic of the chapter. At this final chapter of the book, you might wonder quite naturally what it all does mean. To be elaborate, do all the chapters of the book add up to something? Taken all

© The Author(s), under exclusive license to Springer Nature Switzerland AG 2021
S. N. Seo, *Climate Change and Economics*,
https://doi.org/10.1007/978-3-030-66680-4_13

conclusions from the previous chapters together, what can we say about the future of the Planet's climate and its inhabitants?

Let me put this question again more rigorously. This book has taken a personal storytelling approach in which the chief issues in each of the twelve preceding chapters are discussed independently, say, for the sake of completeness of each chapter. The opposite of this approach is often called the integrated assessment model (IAM). In the IAM approach, all individual topics are inserted into the IAM model as inter-relational entities, from which a single global policy variable is defined and predicted (Nordhaus 1994). It offers a single medicine for the Planet's great illness!

At the finale of this book, however, it is more than pertinent to appreciate the magical solution offered by a host of the integrated assessment models, which is the price of carbon. It is more often than not referred to as carbon tax. It is also conceptualized by many as the social cost of carbon. To be simple, all the actions and changes described in the entirety of this book would add up to something which can be expressed summarily as the price of carbon. From this book's standpoint, as you will soon realize, the carbon price will offer a valuable reference point against which all the actions and changes on climate change can be evaluated.

In the second part of this chapter, having clarified the concept of carbon price in the first part, the present author will come back to analyze the core problem, relying on the carbon tax apparatus, that this book has set out to address, among other things: inter-generational issues and gaps in the debates of climate change. There are quite a few, as will be made clear shortly, critical questions regarding the inter-generational issues that need to be looked into, many of which have been largely set aside by climate researchers for varied reasons. To me, this void of crucial knowledge appears increasingly immense often when I consider many contemporary social happenings and upheavals around the Planet.

One of the wide-spread and popular arguments with regard to the primary topics we are dealing with in this chapter is that future generations or young people will be disproportionately harmed by the global climate change than present generations. Another argument is that younger generations are more willing to make sacrifices than older generations for the Planet's climate. There are many other points of debate that pertain to inter-generational gaps or equities (Yale Communications 2020).

This chapter will provide a careful evaluation of these points, which is urgently called for, in my judgment, in the absence of a rigorous analysis and considering the forcefulness of a popular but untested claim often witnessed. The evaluations will unveil that, contrary to the popular belief, global warming's economic problems and policy decisions are founded on the rationale of inter-generational burden sharing in which one generation's interests are weighed against another generation's interests in a "balanced" way (Arrow et al. 1996; Nordhaus 2007).

In Sect. 13.5, the present author will direct you to another conspicuous gap in global warming discussions that lies between you and the Planet. This is the national gap. To be specific, you will easily notice that each nation is behaving on its own for the sake of the nation while the Planet seems to be behaving on its own for the sake of the Planet. The national gap is where every nation thinks differently in an independent manner without a common goal for the sake of the Planet.

This phenomenon has been salient for the past 30 years in climate policies and negotiations (Seo 2015, 2017). Without rebuilding the bridges among the nations, it will be difficult to connect your individual nation's actions to the Planet's benefit. More importantly, the inter-national gaps will unravel and further elucidate the inter-generational conundrums that I described in the above. In other words, the former, the inter-national gaps, offers another angle for the analysis of the latter, the inter-generational conundrums.

In the final section of this chapter which is also of this book, the present author will attempt to look into the future and elaborate possible paths or pursuits for generation-Zers who are at the moment deeply engaged in the study of climate change and activism. My purpose of doing this is to explain many and diverse paths a young person may wish to cultivate with the afore-described complexities in mind and with a far-sighted planning horizon. For this, the present author will take into account a wide range of possibilities to which global citizens' concerns on global climate change at the present time may or may not dwindle down in the long timeframe beyond the subjects of fierce debate in the nearer future.

I should emphasize, though, that this will be neither a kind of routine career advice nor a prediction of the future. This is rather an attempt by the present author to envision a future evolution of the fields of climate change, economics, and policy negotiations, which is hoped to be a useful guide for young readers. Will the football bounce unexpectedly and unpredictably, say, of climate future? You might wonder. I would too and get well prepared for such slime possibilities some of which will be described as well in this chapter.

13.2   THE PRICE OF CARBON

The price of carbon dioxide (or carbon) is measured as a dollar amount per ton of carbon dioxide (or carbon). Let's imagine a paper permit which allows you to release one ton of carbon dioxide without a penalty. Whenever you release a ton of carbon dioxide, you will have to purchase a permit. The price you pay for the permit is the price of carbon dioxide. Let's suppose that you are allowed to sell the permit to another person. How much would you get from the sale of the permit? The price you receive is the price of carbon dioxide.

If a carbon price system is set up and implemented across the globe or across your country, such as the Canada's carbon pollution pricing scheme, each polluter will have to pay a government-set price for the release of each ton of carbon dioxide (DOJ 2018). For such transactions, the government will most likely set up a carbon tax system within the national tax system through which each polluter is required to pay the carbon tax annually corresponding to the total amount of carbon dioxide emissions.

The price of carbon dioxide would be "universal," that is, harmonized across the Planet if the carbon price system is implemented across the globe. As such, the price of carbon dioxide is also referred to as the social cost of carbon dioxide. There is only one price of carbon dioxide across the society, say, the global community.

The carbon price, or equivalently carbon tax, system was invented by William Nordhaus in the early 1990s, for which the Nobel Memorial Prize in Economics was awarded in 2018 (Nordhaus 2018). The carbon price is a magical medicine offered by the economics profession and indeed a crowning achievement in the field of climate change economics and policy.

With the carbon price system implemented across the Planet, every emitter of carbon dioxide and other greenhouse gases will be forced to pay the single carbon price, through which every emitter will be forced to recognize the financial benefit of reducing or capturing the emissions of carbon dioxide. If designed in a socially optimal manner, the carbon price system will put the global community in a position to address the problem of global warming efficiently, that is, at the lowest cost possible (Nordhaus 1994).

The primary reason that I am introducing the concept of carbon price at this point in the book is, as stated before, to explain to you the concept of an integrated assessment model (IAM). Put differently, I hope to tell at this point how all the elements and results introduced in the preceding chapters would add up to be a meaningful outcome for tackling global climate challenges. To be even more specific, if we were to put all the results from the preceding twelve chapters into the IAM, we will be offered a single price of carbon dioxide today.

How is that possible? The DICE model, short for the Dynamic Integrated Climate and Economy model which is the best documented and most referenced of all the climate change IAMs, traces the life of each ton of carbon dioxide emissions from end to end: from the tailpipes and smokestacks, to the atmosphere, the oceans, to the global climate system, to the ecosystems, and to anthropogenic/economic activities. Further, the DICE model is a dynamic model, meaning that it allows the global community to address the problem of global warming with a long-term planning horizon, say, 300 years from today in a socially welfare optimizing fashion. As such, nearly all the elements and outcomes explained in the preceding twelve chapters are inputs into the DICE model.

Then, how does the DICE model determine the level of carbon dioxide price today and at a future period? In the dynamic global welfare optimization framework, the carbon dioxide price is determined at the socially optimal level of abatement of carbon dioxide at each time period. The carbon price is the marginal cost of abatement at the socially optimal abatement level, at which level the global economic damage of a ton of carbon dioxide is equal to the marginal cost of abatement. Stated differently, the carbon dioxide price is the economic damage incurred by a release of a ton of carbon dioxide today on the global community over the future time periods.

What is the exact level of carbon dioxide price today and at which level will it be a decade later, two decades later, and so on? In Fig. 13.1, I draw a trajectory of carbon price through the twenty-first century which is harmonized across the Planet, which I emphasize is one of the many reported in the literature (Seo 2020). The carbon price for today, that is, for the 2020s stands at 60 US$ per ton of carbon, which is projected to rise throughout the century. By the 2050s, the carbon price stands at 120

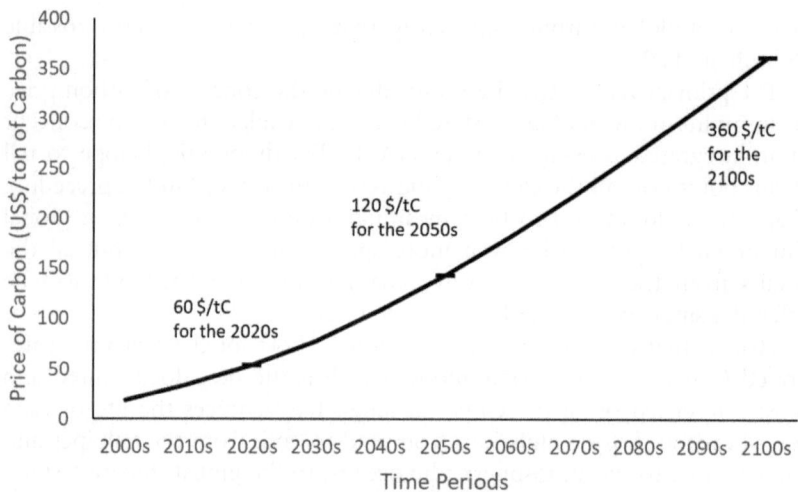

Fig. 13.1 A globally harmonized carbon price trajectory

US$ per ton of carbon. By the end of the century, that is, the 2100s, the harmonized carbon price stands at 360 US$ per ton of carbon.

13.3 GENERATIONAL GAPS

Generational gaps are often exposed and sometimes amplified greatly in the climate change opinions and debates. The gaps across the generations will have a profound implication when the global community has to make a very long-term policy decision in which multiple generations should be accounted for in a fair and rational manner. Regrettably, this elephant in the climate fora has not been clearly addressed by climate researchers while many claims concerning the generational issues come to pass without careful validations. I hope to explain why presently.

The generational gap between two generations arises from manifold aspects of climate change, of which I will draw your attention to the two most prominent. First, it can point to the difference in the perceptions of the two generations about climate change, including the difference in the willingness to make a financial sacrifice for a climate change regulation. The second aspect is the disparate burden of climate change on the two generations. This is often expressed by the statement that the future generations will be the most hurt by the carbon emissions by the present generations.

To appreciate the first type of the generation gap, I will direct you to the results from the on-going annual "Climate Change in the American Mind" survey conducted by the Yale Program on Climate Change Communication (YPCCC) most recently in 2019 (Ballew et al. 2020). The survey relies on the classification of the three generational cohorts: millennial or younger (aged 18–38); generation-X (aged 39–54); baby boomer or older (aged 55+). The gen-Zers, which I often referred to in this book, are belonging to the younger age group within the first cohort, specifically, aged 22 or younger at the time of the survey.

In Table 13.1, I summarize relevant and particularly interesting results from the survey. In response to the two perception questions, millennials and gen-Zers replied with "yes" by about 10% more often than the other generation cohorts did. Specifically, 13% more millennials and gen-Zers are willing to support climate activists and 9% more people from the cohort are willing to identify with climate activists.

Also interestingly, there is no significant difference in the perceptions of generation-X and baby boomer. About a half of the survey respondents (49%) said that they would support climate activists to a degree or

Table 13.1 A survey on generational gaps

Questions	Millennials and Generation-Z (aged 18–38)	Generation-X (aged 39–54)	Baby Boomer (aged +55)
Perception			
Strongly or somewhat support climate activists	62%	49%	49%
Identify with climate activists a great deal or a moderate amount	45%	36%	32%
Action (definitely or probably take the following actions regarding global warming)			
Vote for a candidate	59%	48%	46%
Donate money to an organization	41%	37%	29%

another. But, a strong majority of the respondents from the two cohorts (64–68%) did not identify themselves with climate activists.

Putting aside the perception differential, are they also more willing to take actions? The bottom panel summarizes the results from the two action questions. The millennials and gen-Zers are about 11% more willing to vote for a political candidate because of the concern on global warming than the other cohorts are. By contrast, only 4% more millennials and gen-Zers are willing to donate money to an organization than generation-X. Only 29% of baby boomers said they would donate to an organization, in contrast to 41% of millennials and gen-Zers.

What lies beneath the generational gap in perception and action on global warming, as amply demonstrated by the YPCCC survey summarized in Table 13.1? A key to the answer to the question lies in the illustration of Fig. 13.1: the social cost of carbon dioxide. Note that the social cost ever ticks up through the twenty-first century from one decade to the next. The trajectory tells us, especially gen-Zers, that the economic damage from global warming falling on the generations of people who will live through the twenty-first century will get ever larger from one generation to the next.

As per the figure, the economic damage of one ton of carbon imposed on the generations that are earning incomes during the 2020s is 60 US$/ton. This increases to 120 US$/ton for the generations that are earning incomes during the 2050s, which further increases to 360 US$/ton for the income-earning generations during the 2100s. This is the aforementioned second aspect of the generational gap explained via the interpretation of the social cost of carbon or the economic damage of one ton of carbon. Simply put, people in 2020 will pay 60 US$/ton while people in 2100 will pay 360 US$/ton.

Although media outlets and climate activists are often witnessed to promulgate the above interpretation of inter-generational disparity, it is a flawed economics. This is because it does not account for the fact that the future generations will most certainly live in a better economic condition. Concretely, the personal income will grow in the coming years and decades. To give you an idea, if the national economy would grow at 2% annually, which is a conservative estimate, the nation's GDP will become six times larger than the today's GDP at the end of a century. Taking into account the economic growth and the increased income to be reaped by future generations, you will find that the inter-generational disparity is by far smaller than what it appears to be in Fig. 13.1.

The reason that the inter-generational disparity with regard to the economic burden of climate change may not be severe, regardless of whether or not a global climate policy framework is introduced, is owing to the burden sharing among the generations. In a global optimal policy design espoused by the DICE and many other policy models, all the successive generations of the Planet are assumed to work in concert to achieve the highest welfare for the Planet, given the climate change constraint on the Planet, even though different generations may not even meet face to face.

An optimal economic policy through the efficient carbon pricing will enable the global community to grow its economic welfare at the highest rate possible while simultaneously addressing global climate challenges. If, on the other hand, an inefficient climate policy were to put the rate of economic growth at zero, the 2100 generations would pay, at the maximum, 6 times more than the 2020 generations because of no income growth. The theory of an inter-generational burden sharing, which is at the heart of our analysis in this chapter, will be elucidated presently in the next section.

13.4   Inter-generational Burden Sharing: Fair or Unfair

The concept of an inter-generational burden sharing (IGBS) can be defined as the family of programs that aims at achieving a global welfare optimization across the different generations in tackling the climate change problems, or alternatively, the family of programs that aims to minimize the cost thereof. In the IGBS framework, all generations are envisioned to work in concert to achieve the Planetary goal. The carbon price policy espoused by the DICE model explained in the above is one exemplary policy rooted on the concept of the inter-generational burden sharing.

Here we can further refine it to the three types of the IGBS: a fair IGBS and an unfair IGBS, the latter of which is again of two types: an unfair-to-future-generation IGBS and an unfair-to-present-generation IGBS. The three types of the IGBS are depicted graphically in Fig. 2 via the three different trajectories of the carbon emission control rate at each time period, that is, required of each generation.

The trajectory labeled the Fair IGBS in Fig. 13.2 which is the thick black line is generated assuming that the optimal carbon tax levels already

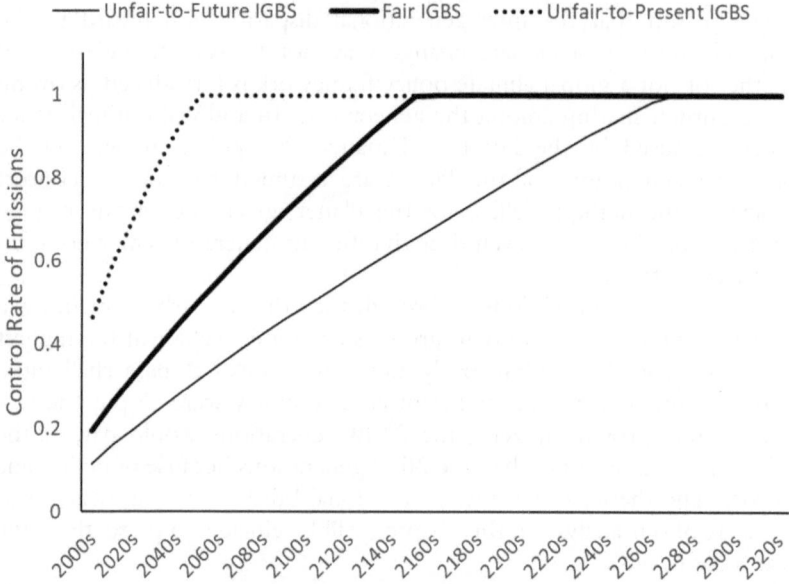

Fig. 13.2 Three types of inter-generational burden sharing via the emission control rate

discussed through Fig. 13.1 will be implemented globally. In fact, for the purpose of expositions in this chapter, I raised the carbon price levels up somewhat by lowering the model's discount rate slightly, the reason for which will be explained shortly. In the Fair IGBS, the global community shall share the burden with the following phased mitigation schedule: Cut greenhouse gas emissions by 25% in 2020, by 50% in 2050, by 75% in 2100, by 100% from the 2150 onwards.

By contrast, in the unfair-to-present IGBS which is the dotted line in the figure, the burden of cutting carbon emissions is shifted to the earlier generations with the following phased schedule: Cut greenhouse gas emissions by 60% in 2020, by 90% in 2050, and by 100% in 2060 onwards. In comparison with the Fair IGBS in the figure, this trajectory of the emission control rate places far greater responsibility on the earlier generations included in the figure, including the present generations.

In the unfair-to-future IGBS which is the thin black line in the figure, the burden of cutting emissions is shifted to the later generations in the figure with the following phased schedule: Cut emissions of greenhouse gases by 15% in 2020, by 30% in 2050, by 45% in 2100, by 60% in 2150, by 80% in 2200, by 90% in 2240, and by 100% in 2260 onwards. This scheme favors the earlier generations by lowering their burden of emissions control, which also means that the later generations will bear the consequence of a higher degree of global warming.

The three types of the IGBS illustrated in Fig. 13.2 on the basis of the carbon mitigation schedule can be explained alternatively through the three corresponding trajectories of the degree of global warming, as is done in Fig. 13.3. The three temperature trajectories are in fact the consequences of adopting the three different schedules of carbon mitigation depicted in Fig. 13.2.

In the temperature trajectory of the Fair IGBS, the global average temperature is predicted to increase by 0.7 degrees Celsius in 2020, to 1.5 °C in the middle of the twenty-first century, to 2.5 °C at the dawn of the twenty-second century. The global average temperature peaks at

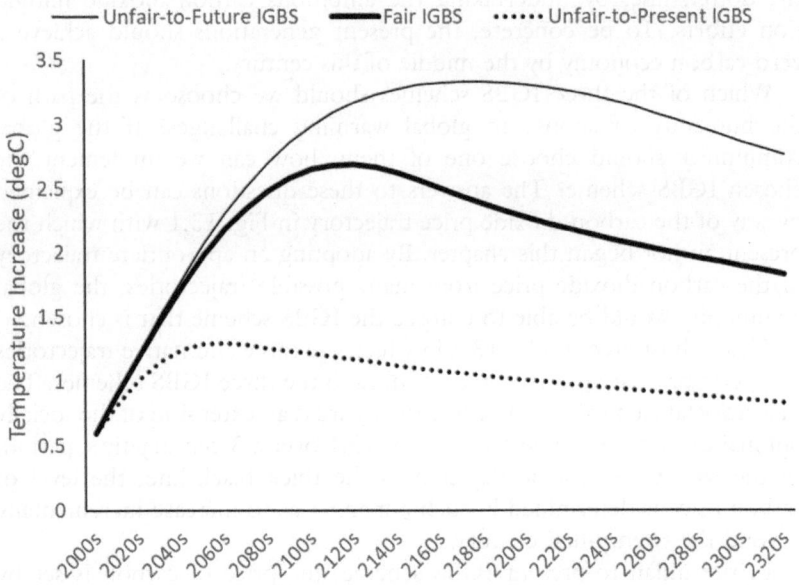

Fig. 13.3 Three types of inter-generational burden sharing *via* the temperature trajectory

around 2.5 °C, after which it starts to come down gradually in the remaining time periods.

In the unfair-to-future IGBS, the global temperature is predicted to keep increasing past the twenty-second century until the end of the twenty-second century. The peak is reached at as high as 3.5 °C. The peak is in fact a plateau that lasts as long as a century. The temperature starts to decline only by the middle of the twenty-third century. In this burden sharing scenario, the future generations of the twenty-second and twenty-third centuries will bear the greater burden of global warming.

In the unfair-to-present IGBS, the global average temperature is predicted to increase and reach the peak at around 1.2 °C as early as the middle of this century. By the middle of the twenty-first century, the temperature starts to fall incrementally through the remaining periods. In this burden sharing scenario, the future generations are spared the burden of a high degree of global warming: To them, the Planetwide warming will be barely noticeable. To accomplish this, though, the present generations should make greater sacrifice, which is illustrated in Fig. 13.2 by the dotted line, by undertaking the ambitious carbon dioxide mitigation efforts. To be concrete, the present generations should achieve a zero-carbon economy by the middle of this century.

Which of the three IGBS schemes should we choose as the path of the humanity's response to global warming challenges? If the global community should choose one of them, how can we implement the chosen IGBS scheme? The answers to these questions can be explained by way of the carbon dioxide price trajectory in Fig. 13.1 with which the present author began this chapter. By adopting an appropriate trajectory of the carbon dioxide price from many possible trajectories, the global community would be able to enforce the IGBS scheme that is chosen.

This is illustrated in Fig. 13.4 in which the three alternative trajectories of carbon price are drawn corresponding to the three IGBS schemes. The trajectory labeled the Fair IGBS in the figure is an extension of the socially optimal carbon price trajectory in Fig. 13.1 over a 3-century time period. In the fair IGBS scheme depicted as the thick black line, the level of carbon price is determined in such a manner as to increase incrementally beyond the twenty-first century.

In the unfair-to-present IGBS scheme, the price of carbon is set by policy-makers at a far higher level starting right from the beginning period compared with that in the far IGBS scheme. The carbon price in 2020 is set at around US$450 per ton of carbon. By the middle of

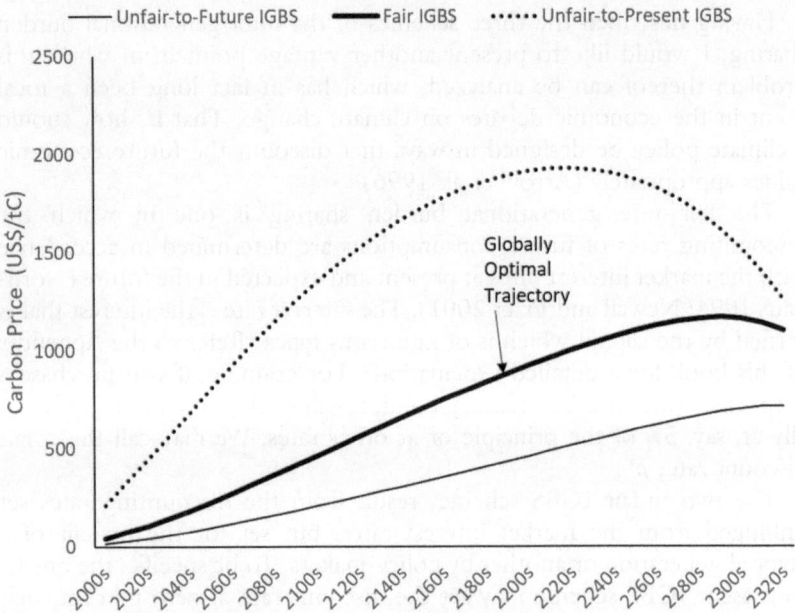

Fig. 13.4 Carbon price and the inter-generational burden sharing schemes (*Note* IGBS=Inter-generational Burden Sharing)

this century, it would hit US$1000 per ton of carbon (/tC). It would reach US$1,500/tC in a century in 2120 and further US$2,000 (/tC) in two centuries in 2220. From then on, the carbon price would fall rather steeply.

In the unfair-to-future IGBS scheme, the price of carbon is set by policy-makers at far lower levels compared with those in the Fair IGBS scheme. It would rise to about US$100 per ton of carbon by the middle of this century, to US$200/tC by the end of this century, to US$460/tC by the end of the twenty-first century. In this burden-sharing scenario, the earlier generations in the figure put off their actions of cutting carbon emissions to the later generations by setting the carbon price deliberately at lower levels. Remind yourself that the lower levels of carbon price set by the present generations (Fig. 13.4) lead to the lower rates of the carbon emission control by the present generations (Fig. 13.2) as well as the higher degrees of global warming for the future generations (Fig. 13.3).

Having described the three schemes of the inter-generational burden sharing, I would like to present another vantage point from which this problem thereof can be analyzed, which has in fact long been a focal point in the economic debates on climate change. That is, how should a climate policy be designed in ways that discount the future economic values appropriately (Arrow et al. 1996)?

The fair inter-generational burden sharing is one in which the discounting rates of future consumptions are determined in accordance with the market interest rates at present and expected in the future (Nordhaus 1994; Newell and Pizer 2001). The interest rate is the interest that is earned by the capital which is of numerous types (Refer to the Appendix of this book for a detailed explanation). For example, if you purchase a corporate bond or a US treasury bond, it will fetch you an interest annually at, say, 5% of the principle or at other rates. We may call this a fair discount rate, ρ^*.

The two unfair IGBS schemes result from the discounting rates set unhinged from the market interest rates, but set for the benefit of a favored generation or another by policy-makers. To be specific, the unfair-to-present IGBS scheme may set the discount rate at zero percent, that is, $\rho = 0$, for the sake of future generations (Weitzman 1998; Stern 2007). The zero discounting rate is unfair to the present generations. Why is it so? This is because the future generation's consumption of US$1 is treated equally (by the policy-makers) to the present generation's consumption of US$1. In the reality of marketplaces, one US dollar in year 2100 is worth only, say, 10 cents in year 2020.

On the other hand, the unfair-to-future IGBS scheme would result from setting the discount rate at the fair discount rate plus α. More concretely, let's say that the discount rate in this scheme is set by near-sighted policy-makers at two times the fair discount rate, that is, $\rho = 2\rho^*$. This is unfair to the future generations because it assumes (in the policy-makers' model) that US$1 in year 2020 will be equal to, say, US$4 in year 2100, although in reality US$1 in year 2020 is worth only, say, US$2 in year 2100. With the discount rate of $2\rho^*$, the value of the consumption by the future generation is given a smaller weight by the policy-makers than it is given under the fair discount rate of ρ^*.

13.5   The Choice: Selfish, Selfless, or Rational?

Having clarified the three schemes of the inter-generational burden sharing, which one should humanity choose as a global policy goal? It should be clear to the readers that only one of the three carbon price trajectories is social welfare optimizing, which is the Fair IGBS. The other trajectories lead to inefficient, that is, non-optimal outcomes for the Planet and its citizens. This is because only one of the three IGBS schemes is a long-term policy that balances the cost with the benefit. One of the balancing acts is the choice of the discount rate consistent with the reality of the marketplaces.

Notwithstanding, the optimality or efficiency is not the only criteria people rely on for their decisions. Faced with this choice of grave consequence regarding inter-generational welfare distributions, I can think of three types of decision-makers: rational, selfish, and selfless. A rational decision-maker would choose the fair IGBS scheme for the social welfare optimality reason just described. A selfish decision-maker would go for the unfair-to-future IGBS scheme because in this scheme the burden of climate change mitigations is lifted off his/her shoulders and imposed on the shoulders of future generations. This is clearly a selfish behavior!

On the contrary, a selfless decision-maker would opt for the unfair-to-present IGBS scheme because in this scheme the burden of climate change mitigations is lifted off the future generations' shoulders and imposed on his/her shoulders. This is again clearly a selfless behavior by the policy-makers who are also among the present generations.

You may be inclined to acting as either a rational decision-maker or a selfless-but-irrational decision-maker. But there is another aspect you must consider before your decision. Let's assume for a moment that you are a selfless decision-maker. The selfless decision-maker may not be the one who will ultimately take costly actions for the sake of the Planet. Why? The selfless policy-maker may just make the choice while the costly actions and sacrifices must be borne by the seven billion individual citizens of the Planet.

As such, the selfless decision-maker is selfless only in reference to the future generations. She/he may be in fact selfish in reference to the individual citizens of the Planet who have to take costly sacrifices. What this simple analysis of being selfless is pointing to is that even the selfless inter-generational burden sharing scheme, which may be in fact your choice, would be extremely difficult to be adopted as a global agreement or treaty. It must be agreed upon by the individual citizens on the Planet at the present time whom will have to bear significantly larger sacrifices.

13.6   BETWEEN YOU AND THE PLANET: NATIONAL GAPS

Before I move on to the next section where I will forward look at the directions to which the literature of climate change and economics is headed, I would like to point out another remarkable gap which has been recognized as a major hurdle in the climate negotiations of the past three decades. If you read through Chapter 12, it will be clear to you that I am referring to the national gap, that is, the disparities among the nations on their commitments (Seo 2017, 2020). In this section, I hope to reinterpret the national gap as another manifestation of the inter-generational problems.

The three carbon price trajectories presented in Fig. 13.4 as well as the emission control trajectories in Fig. 13.2 can be reinterpreted to make clear the international gaps in climate debates. The fair IGBS trajectory is again the trajectory from a globally optimal policy. Instead of the present and future generations, we can interpret the three policy trajectories from the viewpoint of the group of poor countries versus the group of rich countries.

The unfair-to-present IGBS trajectory such as the dotted line in Fig. 13.2 can be interpreted as being unfair to the group of poor but fast-growing countries at the present time. This is because under this treaty the poor but fast-growing countries will be forced to impose a very high carbon price or a very steep emission cut despite the group's priority in economic growth. The high carbon price will essentially prevent the group from, inter alia, building fossil fuel-fired powerplants, installing air-conditioners in homes and factories, and driving automobiles powered by an internal combustion engine. This policy trajectory will be more advantageous to the rich but slow-growing economies.

The unfair-to-future IGBS trajectory, on the other hand, can be interpreted as being unfair to the group of rich but slow-growing countries. This is because this policy trajectory calls for only mild emission cuts during the first half of the twenty-first century, which would give comparative advantages to the poor but fast-growing economies at present. Under this policy trajectory, the world economy will grow substantially each year and as a consequence the world carbon emissions will also grow substantially each year. The economic burden of climate change is put off to the future time periods.

Which policy path should the global community opt for? In making this decision, the climate policy panel of the Planet will face the same

difficult question as described above through the inter-generational decision: Should the global community choose a rational global policy, a
selfish global policy, or a selfless global policy? But, from this viewpoint of
the rich and the poor, the tensions among the negotiators would be far
greater. Directly speaking, can the poor countries sacrifice more for the
sake of the Planet? Can the rich countries sacrifice more for the benefit of
Earth?

13.7 What the Future Holds for Young Climate Activists

I believe most of you have engaged yourself in climate activism from a
very young age, say, as a teenager. I believe you were motivated by a
genuine concern about the Planet, future generations, ecological systems,
the poor and malnourished in developing countries, and extinctions of
animal and fish species. I am also certain that you had many other things
to worry about than climate change, to mention a few, school works,
taking a college entrance examination, taking a part-time job for the
tuition or living costs, social networking with friends, community/family
services, et cetera.

As you grow older and pass the life's milestones one after another,
e.g., becoming an adult, entering a college, graduating from a college,
taking the first job, and getting married, how could you gauge and reeval-
uate your involvements in climate activism, so that you could continue to
contribute to the climate debates while at the same time fully embracing
your new stages of life? I am confident that you the young readers will
face this question once at one point in your life. Before I close this book,
I feel it opportune and right for me, to help your decision, to offer a
rough sketch of the directions and practical issues to which the climate
literature and policy decisions are headed.

First of all, I would encourage you wholeheartedly to keep pursuing
your commitment to the Planet's health. Earth is the only planet that
humans inhabit and as such one of the most valuable resources to the
human race. Having said that, I would also recommend that you should
broaden your areas of concern from global warming to a host of other
challenges that are gravely important to the Planet's well-being. These
challenges are often referred to as global public goods or globally-shared
goods.

A globally-shared good is a good or service that is consumed by the citizens of the Planet collectively. The family of globally-shared goods includes, most notably, global warming, nuclear disarmaments, a pandemic such as the novel coronavirus, a Planetary risk posed by asteroids, a high-risk physics experiment such as the God particle experiment, and artificial intelligence (Seo 2020). The more diverse your areas of activism or scholarship are, the more resourceful you will become as an up and coming generation. Further, you will acquire the additional capacity of weighing one planetary risk against another in your decision.

Second, as you transition from one milestone of life to another during your 20s, 30s, and 40s, you will certainly run into the reality of economic resources. You will realize that you need to earn a living but also that you need monetary resources to achieve many things that you have set out to do. The same can be said of your climate activism. You will need to learn how to balance your activism against the economic concerns of the individuals and societies with which your activism is concerned.

This book was conceived and has been written with a focus on highlighting the economic values against the climate (environmental) concerns you have as well as your life's numerous values. I hope the framework and many stories presented in this book will become ever more meaningful to you as you expand your horizon of interests and activism. I felt recently that this point on economic reality is getting across to even the most radical environmental activists, albeit incrementally.

The third important lesson that this book may deliver to your vision and planning for the future is that you'd better open your eyes to the small things in life and the Planet. By small things, I mean microbehaviors, that is, individuals' choices faced with the reality of a climatic shift. As soon as you start to attempt to open your eyes to tiny things, you will realize that individuals in different locations of the Planet, with different businesses, with different heritages are facing often starkly different choices. You will soon learn to respect such differences.

To elaborate the above, you will soon feel the need to open your eyes to different continents, countries, geographies, businesses, cultural groups, politics, and livelihood systems if you wish to come to grips with the true nature of the globally-shared good that you are concerned with such as global climate change.

This is a far more sophisticated attitude in your analysis than a blanket attitude which is regrettably quite common in the literature and politics of global warming. To mention just one example, it is not hard to find a blanket study and statement that poor countries will, without exception, be severely hurt from global warming. But, as is amply illustrated throughout this book, there are manifold and large differences in many aspects of climate change across the poor countries. In this spirit, I called this book multiple times a big mosaic. Have you noticed it?

The fourth point as per the future direction of the literature and policy decisions is that you will need to awaken yourself to the concept of a "long-term time horizon" in addressing the conundrum of global warming. As a young climate activist, the mention of a "long-term time horizon" may have felt to you like a facile excuse by an old generationer who is trenched in the old ways of life, among other things, a gas-guzzler's life. To the contrary, my experience tells me that some problems could be better addressed with a far-sighted planning horizon than with either a short-sighted planning horizon or a one-shot-and-done attitude. It is especially the case for the international efforts to tackle climate change in which over 200 countries must be involved, which is often missed out by the present negotiators and policy-makers.

The perspective of a long-term planning horizon is in fact unavoidable when it comes to the humanity's interactions with the Planet's climate system. To understand the reality of global warming, to begin with, we are forced to examine at least the past 120 years of climate history since the beginning of the twentieth century, sometimes a million years in the past, or even sometimes many hundreds of millions of years up to the Paleozoic era. The residence time of carbon dioxide in the atmosphere is, for most CO_2 compounds, as long as a century. Further, many climate phenomena of the Planet take many decades to complete a single cycle, for example, the Atlantic Multidecadal Oscillation (AMO), the Pacific Decadal Oscillation (PDO), the Thermohaline Circulation (THC), or the sunspot and solar activity cycles (Le Treut et al. 2007).

When proper attention is paid to these essential characteristics of the Planet's climate system, there would be no solid rationale for the global community to push forcefully to have a complete control over the global climate system in a rushed manner over a short period of time. The popular claims such as "the world is going to end in 12 years," or "end immediately all fossil fuel investments," or "freeze all economic progress" should be viewed as lacking a defensible rationale as well as a strong scientific ground (*USA Today* 2019; CBS News 2020). The salient nature of the Planet's climate system forces us to manage it with a century-long planning horizon in which economic consequences are carefully weighed against the climate challenges over many generations of people who will live and pass on Earth.

With this, I conclude this book. I wish you from the bottom of my heart all the success in your every endeavor.

13.8   Chapter Highlights

- This final chapter offers a grand finale of the book with a set of analyses on generational gaps and the inter-generational burden sharing.
- An integrated assessment model of climate change puts together all the components of climate debates and offers a core policy variable: the price of carbon.
- Two features of the generational gaps are explained: first, millennials and gen-Zers are more willing to support climate activism and act politically; second, future generations will bear the cost of an inaction on global warming.
- The inter-generational burden sharing (IGBS) is explained to be of three types: a fair IGBS, an-unfair-to-future IGBS, and an unfair-to-present-IGBS. The three schemes are differentiated by the trajectories of the emission control rate, of the carbon price, and of the global average temperature increase.
- The chapter concludes with the description of the four key directions of practical importance of the climate literature and policy decisions: globally-shared goods and experiences, economic reality of climate change, micro-actions, and a far-sighted planning horizon.

References

Arrow, Kenneth J., K.G. William Cline, Mohan Munasinghe Maler, R. Squitieri, and Joseph Stiglitz. 1996. Intertemporal Equity, Discounting, and Economic Efficiency. In *Climate Change 1995: Economic and Social Dimensions of Climate Change*, ed. J.P. Bruce, H. Lee, and E.F. Haites. Cambridge, UK: Cambridge University Press.

Ballew, M., J. Marlon, J. Kotcher, E. Maibach, S. Rosenthal, P. Bergquist, A. Gustafson, M. Goldberg, and A. Leiserowitz. 2020. *Young Adults, across Party Lines, are More Willing to Take Climate Action*. New Haven, CT: Yale Program on Climate Change Communication.

CBS News. 2020. *Greta Thunberg Calls for End to All Fossil Fuel Investment "Now" at Davos Forum*. New York, NY: CBS News. Published on January 22, 2020.

Department of Justice (DOJ). 2018. Greenhouse Gas Pollution Pricing Act. S.C. 2018, c. 12, s. 186. DOJ, Canada. Accessed from https://laws-lois.justice.gc.ca/eng/acts/G-11.55/page-1.html.

Le Treut, H., R. Somerville, U. Cubasch, Y. Ding, C. Mauritzen, A. Mokssit, et al. 2007. Historical Overview of Climate Change. In *Climate Change 2007: The Physical Science Basis*, ed. S. Solomon et al. Cambridge: Cambridge University Press.

Newell, Richard, and William Pizer. 2001. *Discounting the Benefits of Climate Change Mitigation: How Much Do Uncertain Rates Increase Valuations?*. Washington, DC: Pew Center on Global Climate Change.

Nordhaus, William. 1994. *Managing the Global Commons*. Cambridge, MA: The MIT Press.

Nordhaus, William. 2007. A Review of the Stern Review on the Economics of Climate Change. *Journal of Economic Literature* 55: 686–702.

Nordhaus, William D. 2018. Climate Change: The Ultimate Challenge for Economics. Prize Lecture. NobelPrize.org. https://www.nobelprize.org/prizes/economic-sciences/2018/nordhaus/lecture/.

Stern, Nicholas. 2007. *The Economics of Climate Change: The Stern Review*. Cambridge: Cambridge University Press.

Seo, S. Niggol. 2015. Helping Low-latitude, Poor Countries with Climate Change. Regulation. Winter 2015–2016, 6–8.

Seo, S. Niggol. 2017. Beyond the Paris Agreement: Climate Change Policy Negotiations and Future Directions. *Regional Science Policy and Practice* 9: 121–40.

Seo, S. Niggol. 2020. *The Economics of Globally Shared and Public Goods*. Amsterdam, NL: Academic Press.

USA Today. 2019. The World is Going to End in 12 Years If We Don't Address Climate Change, Ocasio-Cortez Says. *USA Today*. Published on January 22, 2019.

Weitzman, Martin L. 1998. Why the Far-distant Future Should Be Discounted at Its Lowest Possible Rate. *Journal of Environmental Economics and Management* 36: 201–8.

Yale Program on Climate Change Communication (YPCCC). 2020. *Climate Change in the American Mind April 2020*. New Haven, CT: Yale University.

Appendix: A Brief Exposition of the Essential Economic Theories Used in this Book

The Appendix provides a brief and plain exposition of the economic theories that are essential for your reading and referred to throughout this book. While you read through the book, I recommend you to refer to these explanations if you wish to get certain points in the book clarified. The theories and models covered in the Appendix are a theory of interest, discounting the future, efficient resource uses, Pareto optimality, public goods, globally-shared goods, land value and rent, Gross Domestic Product (GDP), Green GDP, risk and uncertainty, saving and futures, and insurance.

A Theory of Interest

An interest rate is the rate of return on financial assets. Let's suppose you own some cash and lend it to a small business or a bank. The amount you earn from the lending, plus the cash lent, is called the return or interest. The rate of return on your investment (lending) is the interest rate. Interest rates are measured most often in percent per year (Samuelson and Nordhaus 2010).

From an economic viewpoint, the interest rate is the price of borrowing or lending money in the financial market. Many different interest rates are observed across different financial assets such as government bonds, corporate bonds, corporate equities, and consumer loans. The interest

© The Editor(s) (if applicable) and The Author(s), under exclusive   245
license to Springer Nature Switzerland AG 2021
S. N. Seo, *Climate Change and Economics*,
https://doi.org/10.1007/978-3-030-66680-4

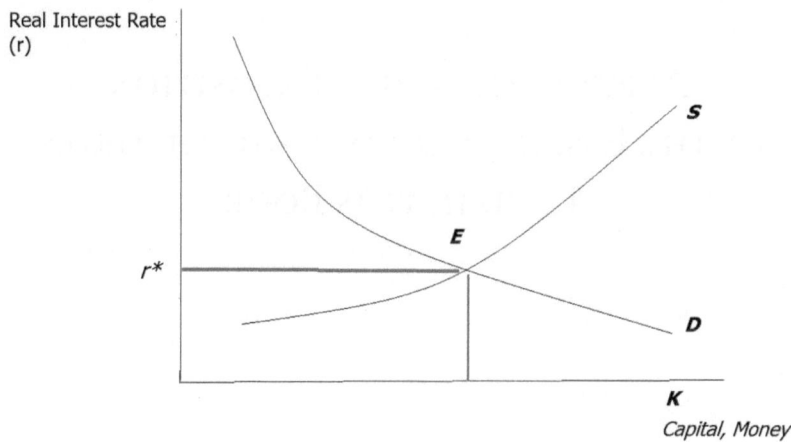

Real Interest Rate (r)

Fig. A.1 Determination of an interest rate

rate varies depending on, among other things, maturity (term), risk, liquidity, and tax treatments of financial assets.

Why are borrowers willing to pay the interest rate for borrowing? This is because the borrowed funds allow them to buy goods and services to satisfy the current consumption needs as well as to make profitable investments in other assets.

In the financial market, the price of money, or equivalently the interest rate, is determined by the interactions between borrowers and lenders (Fisher 1930). As depicted in Fig. A.1, the supply of capital (S) will increase as the interest rate gets higher. On the other hand, the demand for the capital (D) will decrease as the interest rate gets higher. At the intersection of the two market forces, E in the figure, the interest rate is determined.

Let the interest rate be γ. Then, the money you would receive from investing one dollar today, which is called the principle, in two years is as follows, with FV short for Future Value, assuming the interest rate is unchanged:

$FV = 1\$ * (1 + \gamma) * (1 + \gamma)$ or more simply,

$$FV = 1\$ * (1 + \gamma)^2 \tag{A.1}$$

Discounting the Future Consumption

Taking for granted the interest rate (γ) in the financial market, you would soon realize that the one dollar promised in two years is not the same in economic value as the one dollar given immediately. This is because the present value (PV) of the promised dollar in two years is not the same as one dollar today.

Discounting is the process of converting future income into an equivalent present value. This process takes a future dollar amount and reduces it by a discount factor that reflects the appropriate interest rate. The rate at which future incomes are discounted is called the discount rate (Samuelson and Nordhaus 2010).

Let ρ be the discount rate. Continuing with the above example, the discount factor is then $1/(1 + \rho)^2$, with which the present value is calculated as follows:

$$PV = \frac{1\$}{(1 + \rho)^2} \tag{A.2}$$

Efficiency in Resource Allocations

Let's consider a natural resource manager in a rural area in Thailand. Let's suppose that he/she owns the land and water resources on which the following natural resource activities can be performed: crop farming, raising farm animals, forestry, fishery, and mining mineral resources. What is the best way to allocate her resources to each of these activities? The best way is to choose a combination of natural resource uses that would give her the highest profit from all possible combinations. This allocation is referred to as the efficient allocation in economics (Seo 2016).

The marker of the efficient resource allocation is thus: At the efficient resource allocation, the marginal net returns, equivalently profits, from different resource uses should be equated. Let the marginal profit from activity j be written as $\Delta \pi_j$ and the efficient allocation be denoted by $*$. Then, the efficiency condition, also called the equilibrium condition, is as follows:

$$\Delta \pi_i^* = \Delta \pi_j^* \forall i, j. \tag{A.3}$$

To comprehend why Eq. A.3 is the equilibrium condition, we need to consider alternative allocations. Let's consider the alternative allocation in which the marginal profit from activity i is greater than the marginal profit from activity j. This is not an equilibrium condition because she can increase her profit by reallocating her resources marginally, that is, by increasing activity i by one unit and decreasing activity j by one unit (Ricardo 1817; von Thunen 1826). This way of analysis is called the marginal analysis in economics.

Why is this portfolio of natural resource uses called efficient by economists? This is because the maximum profits that can be earned from all other portfolios are smaller than that earned from the efficient portfolio. There will be a profit loss that results from inefficient resource uses. Put differently, an inefficient portfolio leads to a non-optimal economic outcome.

PARETO OPTIMALITY OR EFFICIENCY

Pareto optimality, or equivalently Pareto efficiency, is defined with reference to the macro-economy. Given the limited resources that the national economy is endowed with, the ultimate goal of the economy is to produce the largest amount of goods and services to satisfy the citizens' wants and needs.

Efficiency in the national economy refers to the state in which the economy produces the largest amount of goods and services, given the scarce resources and given the state-of-the-art technology. Note the similarity between this definition and the efficiency in natural resource uses defined in the above.

The efficiency in the macro-economy is best defined by Pareto efficiency after Italian economist Vilfredo Pareto. Let's suppose that there are I individuals in the economy, that is, $i = 1, 2, \ldots, I$. The economy is Pareto efficient when the economy cannot make individual i better off economically without making some other individuals worse off economically at the same time (Pareto 1906). Put differently, the macro-economy is at a Pareto inefficient allocation if the economy can improve individual i's economic welfare without worsening other individuals' welfares by reallocating resources from the inefficient one.

The Pareto efficiency is illustrated in Fig. A.2 with a simple two-person economy. There are two individuals: a and b. In addition, there are two goods produced in this economy: x and y. The individual a's utility functions (U_a1, U_a2, U_a3, U_a4) are drawn from the bottom-left corner

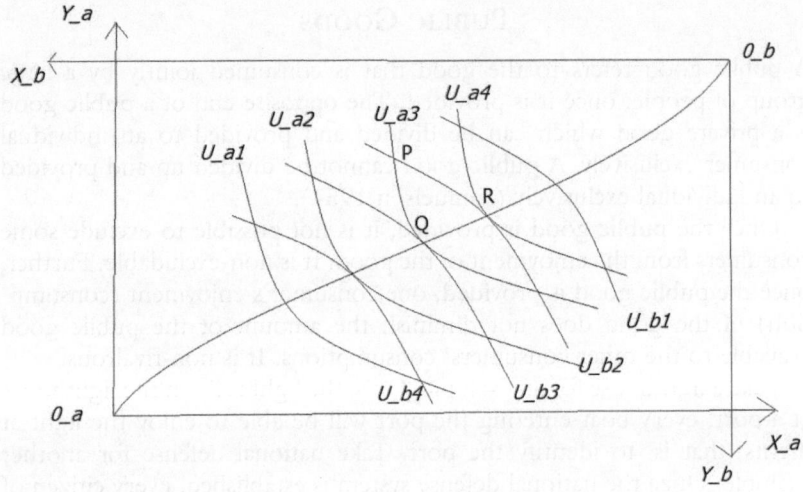

Fig. A.2 Pareto efficiency

of the box (0_a) while the individual b's utility functions (U_b1, U_b2, U_b3, U_b4) are drawn from the top-right corner of the box (0_b).

Let's consider the three allocations of the total amounts of x and y produced in this economy: P, Q, R in the box. At allocation P, the society can reallocate the goods to move to allocation Q. With this reallocation, the society makes individual b better off while making individual a's utility unchanged. Therefore, the allocation P is not Pareto efficient. Similarly, at allocation P, the society can reallocate the goods to move to allocation R, in which case the utility of individual a is improved while the utility of individual b is unchanged. Again, allocation P is not Pareto efficient.

On the other hand, at allocation Q, the society cannot reallocate the goods to make such a Pareto improvement. The same is true of allocation R. To be specific, let's consider reallocation of the goods from allocation Q to allocation R. Although this reallocation improves the welfare of individual a, it worsens the welfare of individual b. The allocation Q is therefore Pareto efficient. The same is true of the allocation R, which you can verify by applying the same analysis.

PUBLIC GOODS

A public good refers to the good that is consumed jointly by a large group of people, once it is provided. The opposite end of a public good is a private good which can be divided and provided to an individual consumer exclusively. A public good cannot be divided up and provided to an individual exclusively (Samuelson 1954).

Once the public good is provided, it is not possible to exclude some consumers from the enjoyment of the good. It is non-excludable. Further, once the public good is provided, one consumer's enjoyment (consumption) of the good does not diminish the amount of the public good available to the other consumers' consumptions. It is non-rivalrous.

Take a lighthouse for an example. Once the light is lit on the lighthouse at a port, every boat entering the port will be able to enjoy the light at nights, that is, to identify the port. Take national defense for another example. Once the national defense system is established, every citizen of the country can enjoy the benefit of it. It is not possible to exclude some boats or some citizens from the consumption of the public goods.

A free-rider problem is inherent as well as prominent in the provision of a public good. Individual consumers tend to wait until others provide the public good and, once provided, they would free ride, that is, enjoy the public good without paying the cost. Since all individuals are expected to act in the same manner, no provision of the public good would result after all.

This point is clarified in Fig. A.3. The figure shows three relationships: a marginal cost function (MC), a marginal benefit function of an individual (MB_i), and a marginal benefit function of the society (MB). The social marginal benefit function is the vertical sum of individual marginal benefit functions. Let's assume, without loss of generality, that the individual marginal benefit functions are the same across the individuals in the society.

Applying the Pareto efficiency rule explained in the above section, the socially optimal (efficient) provision is determined at the intersection of the social marginal benefit function (MB) and the social marginal cost function (MC), which is marked in Fig. A.3. However, a free-riding incentive of individuals leads to an inefficient provision, more specifically, no provision of the public good in the market. Let's suppose that a single individual with the highest marginal benefit function is determined to

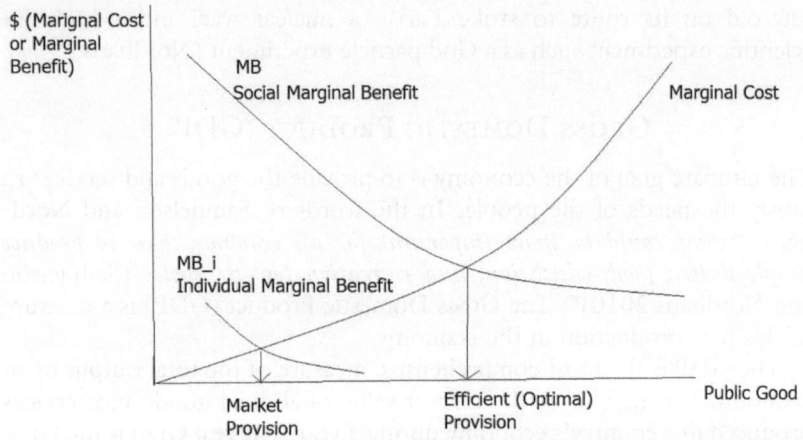

Fig. A.3 An inefficient provision of a public good

provide the public good in the market. In this case, the market provision will be determined at the point where MC is equated to MB_i in the figure. This provision deviates from the Pareto efficient provision. In summary, owing to the free-riding incentive, the public good will be provided inefficiently.

GLOBALLY-SHARED GOODS

Of the range of the public goods explained in the above section, a lighthouse is a local public good while a national defense system is a national public good. To elaborate it, the neighborhood where joint consumption occurs is a locality for the lighthouse while the neighborhood where joint consumption occurs is a nation for the national defense.

Extending this definition, we can think of a public good whose neighborhood of joint consumption is the entire Planet. Such goods are referred to as a global public good or, more recently, as a globally-shared good. The term, a globally-shared good, highlights the salient characteristic of these goods: there is no public sector or government at the Planet level (Seo 2020).

The prominent example of a globally-shared good is the Planet's climate system, which is the global challenge addressed in this book. Other examples are a global pandemic such as COVID-19, a large asteroid on its route to strike Earth, a nuclear war, and a high-risk scientific experiment such as a God particle experiment (Nordhaus 1994).

Gross Domestic Product (GDP)

The ultimate goal of the economy is to provide the goods and services to satisfy the needs of the people. In the words of Samuelson and Nordhaus, *"what could be more important for an economy than to produce ample shelter, food, education, and recreation for its people?* (Samuelson and Nordhaus 2010)". The Gross Domestic Product (GDP) is a measure of this new production in the economy.

The GDP is the most comprehensive measure of the total output of an economy. It is defined as the market value of all final goods and services produced in a country's economy during a year. The real GDP is the GDP adjusted for the inflation, that is, price changes.

Most intuitively, the GDP can be measured as the sum of all incomes (y) earned by the factors of production ($i = 1, 2, \ldots, I$) in the economy, including wages paid to labor, rents earned from lending capital, earnings from financial assets:

$$GDP = \sum_{i=1}^{I} y_i \text{ where } i = 1, 2, \ldots I. \qquad (A.4)$$

A more common method to measure the GDP is the sum of expenditures made by economic sectors:

$$GDP = C + I + G + NX. \qquad (A.5)$$

In the above, C is consumption expenditure, I is investment, G is government expenditure, and NX is the net export, that is, the value of exports minus the value of imports.

Green GDP

Having comprehended the concept of the GDP, you, as a young climate activist, might wonder thus: What if the productions of goods and services, which are captured by the GDP measure, would result in a variety of pollutants which would end up harming human health, capital values, as well as ecosystem health? Such pollutants may be released from burning fossil-fuels for energy generations: to mention some, sulfur dioxide (SO2), ozone (O3), nitrogen oxides (NOx), carbon dioxide (CO2), and particulate matter (PMx).

Economists proposed a Green GDP as an alternative measure to account for such harmful side-effects of production activities (Muller et al. 2011). Let such harmful effects from numerous pollutants be summed up to be D in dollar terms. Then, the green GDP is defined as the following difference:

$$GreenGDP = GDP - D. \qquad (A.6)$$

LAND VALUE AND RENT

To explain numerous economic decisions made by a natural resource manager, the concepts of land rent and land value are essential. Let's suppose a rural farmer in Thailand owns and manages five hectares of farmlands. As explained in the above, he/she will manage her lands to maximize the profit earned from the lands. Let the profit from her lands earned in year t be summed to be π_t. The profit is also referred to as the land rent.

Note that the land rent is also the net income earned from the land or the net revenue. To be more concrete, the land rent is the gross income minus the total cost of managing the lands (Ricardo 1817). Equivalently, the land rent is the gross revenue minus the total cost. Then, the yearly rents can be written as follows:

$$\text{Land rents} = \{\pi_{t+0}, \pi_{t+1}, \pi_{t+2}, \ldots, \pi_{t+k}, \ldots\} \qquad (A.7)$$

From the stream of land rents generated from her land, we can assess the value of the land. You would ask the following as a potential buyer: Knowing that the land will generate net incomes in future periods as shown in Eq. A.7, how much am I willing to pay for the land? You might think at first that the land value may be the sum of all the rents generated in the land over future periods. Thinking twice, you will realize that the future incomes should be discounted to compare them with the present income, which I explained in section "Discounting the Future Consumption" of this Appendix. After incorporating the discounting of the future stream of yearly rents with the discount rate γ, the land value is defined as follows (Seo 2016):

$$\text{Land value} = \sum_{k=0}^{\infty} \frac{\pi_{t+k}}{(1+\gamma)^k}. \qquad (A.8)$$

RISK AND UNCERTAINTY

Life is full of uncertainty, so is any economic investment decision. The larger the uncertainty, the higher the risk. The smaller the uncertainty, the lower the risk. Some investments are riskier than other investments, so are some natural events than others. Some investors are risk takers while others are not.

Let's consider again a farmer in Thailand cultivating a hectare of rice land. At the time of planning, the annual yield of rice at the end of the harvesting seasons is uncertain. It depends upon many factors including the weather of the year, the region's climate regime, soils, labor availability, national policy changes, and international changes. The yield risk at the farmland can be defined as the variability of the annual rice yield across the years at the farmland. Statistically, it can be measured by the standard deviation (SD) of yearly yields or the coefficient of variation (CV) of yearly yields. The latter is a measure of variability relative to the average yield.

In the same manner, the revenue risk faced by the famer can be defined. In other words, it is the degree of variability across the years of yearly net revenues earned from the sales of rice harvests. It can be defined using either SD or CV of yearly net revenues.

Individuals, including Thai farmers, are observed to exhibit different attitudes toward the risk. In this regard, three types of individuals or investors can be defined: risk-averse, risk neutral, and risk-loving (Arrow 1971). In plain terms, a risk-loving investor would take a high return portfolio despite the portfolio's high risk. A risk-averse investor, on the other hand, would take a low risk portfolio despite the portfolio's low return.

An efficient decision varies across the individuals because of their attitudes toward the risk. I can explain this with the efficient portfolio frontier shown in Fig. A.4. The vertical axis is the expected return while the horizontal axis is the standard deviation of the return. As such, each portfolio in the figure reveals a risk and return tradeoff of an individual investor.

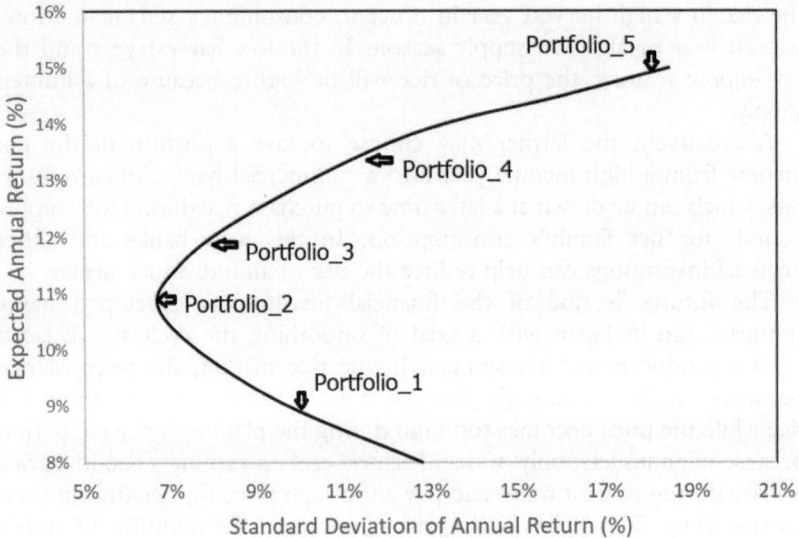

Fig. A.4 Risk and return in the efficient portfolio frontier

On the frontier, five portfolios are marked. A risk-loving investor would choose Porfolio_5, a high return portfolio, despite high risk. A risk-averse investor would choose Portfolio_2, a low risk portfolio, despite low expected return.

Let's compare the choice of Portfolio_3 with the choice of Portfolio_2. A risk neutral investor may switch from Portfolio_2 to Portfolio_3 because Portfolio_3 gives a higher return despite a marginal increase in the risk than Portfolio_2. Let's this time consider Portfolio_1. A rational investor would not choose this portfolio because it yields both a lower return and a higher risk than Portfolio 2. An extended analysis of the efficient portfolio frontier results ultimately in the capital asset pricing theory (Sharpe 1964).

MANAGING THE RISK: SAVING AND FUTURES

Let alone the Thai rice farmer, all managers of natural resources and financial assets must face the risk. If the risk is modest, they can cope with some of the risk on their own. For example, the Thai farmer may store the rice in a high harvest year in order to consume (or sell) it in a low-harvest year or at a low-supply season. In the low-harvest year and the low-supply seasons, the price of rice will be higher because of a limited supply.

Alternatively, the farmer may choose to save a portion of the net income from a high-income year into a commercial bank and earn interests, which can be drawn at a later time to purchase rice during low-supply periods for her family's consumption. In this way, banks and other financial institutions can help reduce the risk of an individual farmer.

The futures is one of the financial instruments developed many centuries ago in Japan with a goal of smoothing the cyclical risk faced by rice producers and consumers. In the rice market, the price of rice becomes too low during the harvest time because of an oversupply of rice while the price becomes too high during the planting-growing period because of an undersupply of rice. Farmers end up earning a too low price for rice during harvest times and pay a too high price for rice during non-harvest times. This is the cyclical risk common in the traditional farming communities (Shiller 2004).

To manage this cyclical risk, a futures market was gradually formed and advanced. A rice farmer would go into a futures contract during the planting season with a futures contract buyer. The futures contract specifies a certain amount of rice to be delivered by the seller to the buyer at a future date at a specified price (Seo 2018).

The futures market improves the economic welfare of both producers and consumers of rice. In other words, it leads to a Pareto improvement of the society. Through the futures contract, the producer can avoid selling the rice at too low prices owing to an oversupply during the harvest times. On the other hand, the buyer can avoid purchasing at too high prices owing to an undersupply.

Developed in the rice market in the medieval Japan, the futures contracts are nowadays traded for agricultural commodities, natural resources, and financial assets. Futures transactions take place through a formalized market such as the Chicago Mercantile Exchange and the New York Stock Exchange.

Managing the Risk: Insurance

An insurance contract is another financial instrument developed in the medieval Europe and designed to help reduce the risk of a rarely occurring catastrophic event. Readers may be familiar with a housefire insurance, an automobile insurance, a life insurance, a flood insurance, among other things, as these products are advertised widely in the media. Notice that the insured events in these insurances are all a very low probability event and simultaneously a very high damage event (Seo 2018).

For an insurance to work, in addition, there should be a very large number of people who are faced with the same risk. The core economic concept of the insurance is pooling the risk, which means that the insurance is a financial instrument that is capable of spreading the risk, also known as pooling the risk, of a disaster across a large number of insured individuals who are faced with the same risk (Shiller 2004).

An insurance contract is structured in the following manner. Once a buyer enters into a housefire insurance contract with an insurer, she agrees to pay an insurance premium monthly throughout the contract period. In return, the insurer guarantees the buyer, that is, the insured, that the insurance payout specified in the contract shall be made in the event that a housefire occurs in the buyer's house (Fabozzi et al. 2009).

In this book, a national flood insurance is explained as an important aspect of managing the catastrophic risk of hurricanes in the US. Another is the crop insurance and subsidy of the US which provides insurance payments in the event of a catastrophic loss in crop yields or crop revenues.

Again, the insurance contract leads to a Pareto improvement of the society, that is, improves the welfare of both the buyers and sellers. I leave it to the readers to figure out why such is the case.

REFERENCES

Arrow, Kenneth J. 1971. *Essays in the Theory of Risk Bearing*. Chicago: Markham Publishing Co.

Fabozzi, Frank, Franco P. Modigliani, and Frank J. Johns. 2009. *Foundations of Financial Markets and Institutions*. New York, NY: Prentice Hall.

Fisher, Irving. 1930. *The Theory of Interest*. New York: Macmillan.

Muller, Nicholas Z., Robert Mendelsohn, and William Nordhaus. 2011. Environmental Accounting for Pollution in the United States. *American Economic Review* 101: 1649–75.

Nordhaus, William. 1994. *Managing the Global Commons*. Cambridge, MA: MIT Press.

Pareto, Vilfredo. 1906. Manual for Political Economy. In *2014*, ed. A. Montesano, A. Zanni, L. Bruni, J.S. Chipman, and M. McLure . Oxford: Oxford University Press.

Ricardo, David. 1817. *On the Principles of Political Economy and Taxation.* London, UK: John Murray.

Samuelson, Paul. 1954. The Pure Theory of Public Expenditure. *The Review of Economics and Statistics* 36: 387–89.

Samuelson, Paul, and William Nordhaus. 2010. *Economics*, 19th ed. New York, NY: McGraw-Hill.

Seo, S. Niggol. 2016. *Microbehavioral Econometric Methods: Theories, Models, and Applications for the Study of Environmental and Natural Resources.* Amsterdam, NL: Academic Press.

Seo, S. Niggol. 2018. *Natural and Man-made Catastrophes: Theories, Economics, and Policy Designs.* Hoboken, NJ: Wiley-Blackwell.

Seo, S. Niggol. 2020. *The Economics of Globally Shared and Public Goods.* Amsterdam, NL: Academic Press.

Sharpe, William F. 1964. Capital Asset Prices: A Theory of Market Equilibrium Under Conditions of Risk. *Journal of Finance* 19 (3): 425–42.

Shiller, Robert J. 2004. *The New Financial Order: Risk in the 21st Century.* Princeton, NJ: Princeton University Press.

von Thunen, J.H. [1966] 1826. *The Isolated State* (CM Wartenberg, trans., 1966). New York, NY: Pergamon Press.

INDEX